# SpringerBriefs in Physics

More information about this series at http://www.springer.com/series/8902

Alexander Streltsov

# Quantum Correlations Beyond Entanglement

and Their Role in Quantum Information Theory

 Springer

Alexander Streltsov
ICFO—The Institute of Photonic Sciences
Barcelona, Castelldefels
Spain

ISSN 2191-5423        ISSN 2191-5431   (electronic)
ISBN 978-3-319-09655-1    ISBN 978-3-319-09656-8   (eBook)
DOI 10.1007/978-3-319-09656-8

Library of Congress Control Number: 2014946381

Springer Cham Heidelberg New York Dordrecht London

Printed on acid-free paper

Springer is part of Springer Science+Business Media (www.springer.com)

# Acknowledgments

Parts of this work are based on the research carried out during my Ph.D. at the Heinrich-Heine-Universität Düsseldorf [1]. I thank Dagmar Bruß for the excellent supervision of the Ph.D., and I also thank the following people for discussion and support: Gerardo Adesso, Remigiusz Augusiak, Otfried Gühne, Hermann Kampermann, Matthias Kleinmann, Maciej Lewenstein, Tobias Moroder, Marco Piani, and Wojciech Zurek. I also acknowledge financial support by the Alexander von Humboldt-Foundation, the John Templeton Foundation, EU IP SIQS, ERC AdG OSYRIS, and EU-Spanish Ministry CHISTERA DIQIP.

## Reference

1. Streltsov, A: The role of quantum correlations beyond entanglement in quantum information theory. Ph.D. thesis, Heinrich-Heine-Universität Düsseldorf. http://docserv.uni-duesseldorf.de/servlets/DocumentServlet?id=26112 2013.

# Contents

**1 Introduction** ............................................. 1
   References ............................................ 3

**2 Quantum Theory** ......................................... 5
   2.1 Quantum States ..................................... 5
   2.2 Quantum Measurements and Operations ................. 6
   2.3 Reduced Density Operator ............................ 7
   2.4 Entropy and Mutual Information ...................... 8
   2.5 Distance Between Density Operators .................. 9
   References ............................................ 10

**3 Quantum Entanglement** ................................... 11
   3.1 Definition ......................................... 11
   3.2 Local Operations and Classical Communication ........ 12
   3.3 Entanglement as a Resource ......................... 12
   3.4 Entanglement Measures .............................. 13
   References ............................................ 15

**4 Quantum Correlations Beyond Entanglement** ............... 17
   4.1 Definition ......................................... 17
   4.2 Measures of Quantum Correlations .................... 18
      4.2.1 Quantum Discord ............................. 18
      4.2.2 General Measures of Quantum Correlations ..... 20
   References ............................................ 22

**5 Quantum Discord in Quantum Information Theory** .......... 23
   5.1 Remote State Preparation ........................... 23
      5.1.1 Deterministic Remote State Preparation ....... 23
      5.1.2 Remote State Preparation in the Presence of Noise ..... 24
      5.1.3 Average Payoff .............................. 26

        5.1.4   The Role of Quantum Correlations . . . . . . . . . . . . . . .   27
        5.1.5   Discussion . . . . . . . . . . . . . . . . . . . . . . . . . . . . . . .   29
   5.2   Entanglement Distribution. . . . . . . . . . . . . . . . . . . . . . . . .   29
        5.2.1   General Protocol for Entanglement Distribution. . . . . . . .   29
        5.2.2   The Role of Quantum Correlations . . . . . . . . . . . . . . .   30
        5.2.3   Discussion . . . . . . . . . . . . . . . . . . . . . . . . . . . . . . .   33
   5.3   Transmission of Correlations. . . . . . . . . . . . . . . . . . . . . . . .   34
        5.3.1   Classical Transmission of Correlations . . . . . . . . . . . . .   34
        5.3.2   The Role of Quantum Correlations . . . . . . . . . . . . . . .   35
        5.3.3   Quantum Transmission of Correlations. . . . . . . . . . . . .   38
        5.3.4   Equivalence of Quantum and Classical Transmission
                for Pure States . . . . . . . . . . . . . . . . . . . . . . . . . . . . .   40
        5.3.5   Discussion . . . . . . . . . . . . . . . . . . . . . . . . . . . . . . .   42
   References . . . . . . . . . . . . . . . . . . . . . . . . . . . . . . . . . . . . . .   42

**6  Outlook** . . . . . . . . . . . . . . . . . . . . . . . . . . . . . . . . . . . . . . . .   45
   References . . . . . . . . . . . . . . . . . . . . . . . . . . . . . . . . . . . . . .   46

**Index** . . . . . . . . . . . . . . . . . . . . . . . . . . . . . . . . . . . . . . . . . . .   47

# Abstract

Quantum entanglement is the most popular kind of quantum correlations, and its fundamental role in several tasks in quantum information theory like quantum cryptography, quantum dense coding, and quantum teleportation is undeniable. However, recent results suggest that various applications in quantum information theory do not require entanglement, and that their performance can be captured by a new type of quantum correlation that goes beyond entanglement. Quantum discord, introduced by Zurek more than a decade ago, is the most popular candidate for such general quantum correlations. In this work we give an introduction to this modern research direction. After a short review of the main concepts of quantum theory and entanglement, we present quantum discord and general quantum correlations, and discuss three applications based on this new type of correlations: remote state preparation, entanglement distribution, and transmission of correlations. We also give an outlook to other research in this direction.

# Chapter 1
# Introduction

Quantum entanglement has fascinated the minds of physicists since the very inception of quantum theory [1]. Entangled quantum systems can behave in a bizarre way, exhibiting features which seem to contradict "our common sense notions of how the world works" [2, p. 114]. This was first pointed out in a seminal work by Einstein, Podolsky, and Rosen, who concluded that the quantum theory must be incomplete [3]. However, about 30 years after Einstein's objection, Bell proposed an experiment, which aimed to distinguish between predictions made by quantum theory on the one hand, and Einstein's arguments on the other hand [4]. Bell's ideas served as a starting point for Clauser, Horne, Shimony, and Holt, who formulated an inequality which is known today as the CHSH inequality [5]. Following Einstein et al., Nature should respect the CHSH inequality, and the fact that it can be violated in quantum theory demonstrates the incompleteness of quantum mechanics.

Due to its simplicity, the CHSH inequality could be tested experimentally by Freedman and Clauser already short time after its discovery [6]. The data showed a violation of the CHSH inequality, thus invalidating Einstein's arguments, in favor of the quantum mechanical description of Nature. Later in the years 1981/1982 Aspect et al. performed three experiments [7–9], confirming the results of Freedman and Clauser. Since that time, several experiments have demonstrated violation of the CHSH inequality, although some loopholes still remained open [10].

The formal definition of entanglement as we use it today can be dated back to the year 1989, when Werner extended the concept of entanglement to all mixed quantum states [11]. Werner's work can be regarded as the starting point for the theory of entanglement, which studies properties and implications of entanglement, and its role in such fundamental tasks like quantum cryptography [12], quantum dense coding [13], and quantum teleportation [14]. Several important contributions to the theory of entanglement also came from the Horodecki family: one example is the discovery of bound entanglement [15]. Bound entangled states need some amount of entanglement to be created, but cannot be used for the extraction of any pure entangled state. A comprehensive review on this topic can be found in [10].

The role of entanglement in quantum algorithms is still subject of extensive debate. This is due to the results by Jozsa and Linden, who showed that a quantum com-

© The Author(s) 2015
A. Streltsov, *Quantum Correlations Beyond Entanglement*,
SpringerBriefs in Physics, DOI 10.1007/978-3-319-09656-8_1

puter operating on a pure state needs entanglement in order to have an exponential speedup compared to classical computation [16, 17]. Although exponential speedup of a quantum computer is not yet rigorously proven, there is strong evidence for its existence. One of the most prominent examples pointing in this direction is Shor's prime factorization algorithm proposed in [18]. The algorithm is able to find the prime factors for any product of two primes on a quantum computer, where the time for the computation grows polynomially in the number of the input bits. This is significantly faster, compared to the best known classical algorithm, which exhibits an exponential increase of the running time.

Due to the presence of entanglement in Shor's algorithm [17] one might be tempted to see entanglement as the key resource for quantum computation. While for *pure state* quantum computation this is indeed the case, the situation becomes more involved if *mixed state* quantum computation is considered [17]. A popular example for mixed state quantum computation has been presented by Knill and Laflamme [19]. Surprisingly, their algorithm is able to solve certain problems efficiently for which no efficient classical algorithm is known even with vanishingly little entanglement [20]. This finding triggered the search for quantum correlations beyond entanglement, which should be responsible for the efficiency of a quantum computer.

*Quantum discord*, introduced by Zurek in the year 2000, has been recognized as a possible candidate for those general quantum correlations [21, 22]. On the one hand, quantum discord can even exist in systems which are not entangled. On the other hand, it has been shown that the algorithm presented by Knill and Laflamme exhibits nonvanishing amount of discord [23]. An even stronger statement has been made by Eastin, who showed that mixed state quantum computation with zero discord in each step can be simulated efficiently on a classical computer [24]. Three years after Zurek has proposed quantum discord as a new kind of quantum correlations beyond entanglement, he gave it an alternative thermodynamical interpretation [25]. He considered the amount of work which can be extracted from a quantum system by a classical and a quantum Maxwell's demon. He showed that the quantum demon is more powerful, since it can operate on the whole quantum state, while the classical demon is restricted to local subsystems only. Zurek concluded that more work can be extracted in the quantum case, and this quantum advantage is related to the quantum discord.

Approximately at the same time when Zurek defined quantum discord, a closely related quantity has been proposed by Henderson and Vedral [26]. The authors aimed to separate correlations into quantum and purely classical parts by postulating several reasonable properties. This approach is significantly different from Zurek's, and the fact that both arrive at the same result is surprising. Another related quantity is the *information deficit*, presented in [27]. The authors study the amount of work, which can be extracted from a heat bath using a mixed quantum state. If the mixed state is shared by two parties, the amount of extractable work is usually smaller, compared to the case where the whole state is in possession of a single party. The difference of these two quantities is the information deficit.

In the light of these results, it is not surprising that in the last few years an enormous amount of research has been devoted to tasks in quantum information theory which are not based on entanglement [28]. Quantum discord has been related to the performance of some of those tasks: remote state preparation [29] and information encoding [30] being popular examples. Experimental techniques for detecting general quantum correlations have also been presented [31]. In this work, we give an introduction to general quantum correlations beyond entanglement and present a detailed discussion on their role for remote state preparation [29], entanglement distribution [32, 33], and transmission of correlations [34, 35]. We start by briefly reviewing the mathematical framework of quantum theory, followed by a short introduction to quantum entanglement. After introducing quantum discord and related quantifiers of quantum correlations, we discuss their role in quantum information theory, and also present a short outlook on other research directions.

# References

1. Schrödinger, E.: Die gegenwärtige Situation in der Quantenmechanik. Naturwissenschaften **23**, 807–812 (1935)
2. Nielsen, M.A., Chuang, I.L.: Quantum Computation and Quantum Information. Cambridge University Press, Cambridge (2000)
3. Einstein, A., Podolsky, B., Rosen, N.: Can Quantum-mechanical description of physical reality be considered complete? Phys. Rev. **47**, 777–780 (1935)
4. Bell, J.S.: On the Einstein-Podolsky-Rosen paradox. Physics (Long Island City, N.Y.) **1**, 195–200 (1964)
5. Clauser, J.F., Horne, M.A., Shimony, A., Holt, R.A.: Proposed experiment to test local hidden-variable theories. Phys. Rev. Lett. **23**, 880–884 (1969)
6. Freedman, S.J., Clauser, J.F.: Experimental test of local hidden-variable theories. Phys. Rev. Lett. **28**, 938–941 (1972)
7. Aspect, A., Grangier, P., Roger, G.: Experimental tests of realistic local theories via Bell's theorem. Phys. Rev. Lett. **47**, 460–463 (1981)
8. Aspect, A., Dalibard, J., Roger, G.: Experimental test of Bell's inequalities using time- varying analyzers. Phys. Rev. Lett. **49**, 1804–1807 (1982)
9. Aspect, A., Grangier, P., Roger, G.: Experimental realization of Einstein-Podolsky-Rosen-Bohm Gedankenexperiment: a new violation of Bell's inequalities. Phys. Rev. Lett. **49**, 91–94 (1982)
10. Horodecki, R., Horodecki, P., Horodecki, M., Horodecki, K.: Quantum entanglement. Rev. Mod. Phys. **81**, 865–942 (2009)
11. Werner, R.F.: Quantum states with Einstein-Podolsky-Rosen correlations admitting a hidden-variable model. Phys. Rev. A **40**, 4277–4281 (1989)
12. Ekert, A.K.: Quantum cryptography based on Bell's theorem. Phys. Rev. Lett. **67**, 661–663 (1991)
13. Bennett, C.H., Wiesner, S.J.: Communication via one- and two-particle operators on Einstein-Podolsky-Rosen states. Phys. Rev. Lett. **69**, 2881–2884 (1992)
14. Bennett, C.H. et al.: Teleporting an unknown quantum state via dual classical and Einstein-Podolsky-Rosen channels. Phys. Rev. Lett. **70**, 1895–1899 (1993)
15. Horodecki, M., Horodecki, P., Horodecki, R.: Mixed-state entanglement and distillation: is there a "Bound" entanglement in nature? Phys. Rev. Lett. **80**, 5239–5242 (1998)
16. Jozsa, R.: Entanglement and Quantum Computation (1997). arXiv:quant-ph/9707034v1

17. Jozsa, R., Linden, N.: On the role of entanglement in quantum-computational speed-up. Proc. R. Soc. A **459**, 2011–2032 (2003)
18. Shor, P.: Algorithms for quantum computation: discrete logarithms and factoring. In: Goldwasser, S. (ed.) Proceedings of the 35th Annual Symposium on the Foundations of Computer Science, 124 (IEEE Computer Society. Los Alamitos, CA (1994))
19. Knill, E., Laflamme, R.: Power of one bit of quantum information. Phys. Rev. Lett. **81**, 5672–5675 (1998)
20. Datta, A., Flammia, S.T., Caves, C.M.: Entanglement and the power of one qubit. Phys. Rev. A **72**, 042316 (2005)
21. Zurek, W.H.: Einselection and decoherence from an information theory perspective. Ann. Phys. (Leipzig) **9**, 855–864 (2000)
22. Ollivier, H., Zurek, W.H.: Quantum discord: a measure of the quantumness of correlations. Phys. Rev. Lett. **88**, 017901 (2001)
23. Datta, A., Shaji, A., Caves, C.M.: Quantum discord and the power of one qubit. Phys. Rev. Lett. **100**, 050502 (2008)
24. Eastin, B.: Simulating Concordant Computations (2010). arXiv:1006.4402v1
25. Zurek, W.H.: Quantum discord and Maxwell's demons. Phys. Rev. A **67**, 012320 (2003)
26. Henderson, L., Vedral, V.: Classical, quantum and total correlations. J. Phys. A: Math. Gen. **34**, 6899 (2001)
27. Oppenheim, J., Horodecki, M., Horodecki, P., Horodecki, R.: Thermodynamical approach to quantifying quantum correlations. Phys. Rev. Lett. **89**, 180402 (2002)
28. Modi, K., Brodutch, A., Cable, H., Paterek, T., Vedral, V.: The classical-quantum boundary for correlations: discord and related measures. Rev. Mod. Phys. **84**, 1655–1707 (2012)
29. Dakić, B. et al.: Quantum discord as resource for remote state preparation. Nat. Phys. **8**, 666–670 (2012)
30. Gu, M. et al.: Observing the operational significance of discord consumption. Nat. Phys. **8**, 671–675 (2012)
31. Gessner, M. et al.: Local detection of quantum correlations with a single trapped ion. Nat. Phys. **10**, 105–109 (2014)
32. Streltsov, A., Kampermann, H., Bruß, D.: Quantum cost for sending entanglement. Phys. Rev. Lett. **108**, 250501 (2012)
33. Chuan, T.K. et al.: Quantum discord bounds the amount of distributed entanglement. Phys. Rev. Lett. **109**, 070501 (2012)
34. Streltsov, A., Zurek, W.H.: Quantum discord cannot be shared. Phys. Rev. Lett. **111**, 040401 (2013)
35. Brandão, F.G. S.L., Piani, M., Horodecki, P.: Quantum Darwinism is Generic (2013). arXiv:1310.8640v1

# Chapter 2
# Quantum Theory

## 2.1 Quantum States

In quantum mechanics, any physical system is completely described by a state vector $|\Psi\rangle$ in a Hilbert space $\mathcal{H}$. A system with a two-dimensional Hilbert space is also called a *qubit* (quantum bit). If not otherwise stated, we consider a Hilbert space with an arbitrary but finite dimension. For two parties, Alice ($A$) and Bob ($B$), with Hilbert spaces $\mathcal{H}_A$ and $\mathcal{H}_B$ the total Hilbert space is a tensor product of the subsystem spaces: $\mathcal{H}_{AB} = \mathcal{H}_A \otimes \mathcal{H}_B$.

Any system which is described by a single state vector is said to be in a *pure state*. However, in a realistic experimental setup the physical state of the considered system is not completely known. If the system is in the pure state $|\psi_i\rangle$ with probability $p_i$, the physical state of the system can be described using the *density operator*

$$\rho = \sum_i p_i |\psi_i\rangle \langle \psi_i| . \tag{2.1}$$

The state of such a system is called *mixed state*. In the following, whenever we talk about quantum states, we usually mean mixed states.

In order to have a meaningful physical interpretation, any density operator has the following two properties:

- $\rho$ has trace equal to one:

$$\mathrm{Tr}[\rho] = 1, \tag{2.2}$$

- $\rho$ is a positive operator:

$$\langle \psi | \rho | \psi \rangle \geq 0 \tag{2.3}$$

for any vector $|\psi\rangle$.

Note that the second property also implies that $\rho$ is Hermitian: $\rho^\dagger = \rho$. These two condition are essential for the definition of quantum measurements and operations, which is presented in the following.

© The Author(s) 2015

A. Streltsov, *Quantum Correlations Beyond Entanglement*,
SpringerBriefs in Physics, DOI 10.1007/978-3-319-09656-8_2

## 2.2 Quantum Measurements and Operations

Quantum measurement is one of the most important concepts in quantum theory. Most physicists are familiar with the *projective measurement*: for a spin-$\frac{1}{2}$ particle in the state

$$|\psi\rangle = a|\uparrow\rangle + b|\downarrow\rangle, \tag{2.4}$$

the probability to measure "spin up" or "spin down" is given by $p(\uparrow) = |a|^2$ or $p(\downarrow) = |b|^2 = 1 - p(\uparrow)$. Moreover, the measurement postulate of quantum mechanics tells us that the quantum state after the measurement is either $|\uparrow\rangle$ or $|\downarrow\rangle$, depending on the outcome of the measurement.

In quantum information theory, a more general definition is considered. A general quantum measurement is described by a collection $\{E_i\}$ of *measurement operators* that satisfy the completeness equation:

$$\sum_i E_i^\dagger E_i = \mathbb{1}, \tag{2.5}$$

where $\mathbb{1}$ is the identity operator. Given a density operator $\rho$ and the set of measurement operators $\{E_i\}$, the probability that the result $i$ occurs is given by

$$p_i = \mathrm{Tr}[E_i^\dagger E_i \rho]. \tag{2.6}$$

After the measurement with outcome $i$, the state of the system is described by the density operator

$$\rho_i = \frac{1}{p_i}(E_i \rho E_i^\dagger). \tag{2.7}$$

The set of operators

$$M_i = E_i^\dagger E_i \tag{2.8}$$

is also called positive operator-valued measure (*POVM*). Due to the completeness Eq. (2.5), the POVM elements $M_i$ sum up to the identity operator: $\sum_i M_i = \mathbb{1}$. Moreover, due to Eq. (2.6) the probabilities $p_i$ can also be obtained from the POVM elements $M_i$: $p_i = \mathrm{Tr}[M_i \rho]$. The positivity of the density operator $\rho$ in Eq. (2.3) implies that all probabilities are nonnegative: $p_i \geq 0$. The completeness Eq. (2.5) together with Eq. (2.2) implies that the probabilities sum up to one: $\sum_i p_i = 1$.

For a projective measurement, the operators $E_i$ are orthogonal projectors: $E_i E_j = \delta_{ij} E_i$. *Von Neumann measurement* is a special type of a projective measurement, where the measurement operators $E_i$ are orthogonal projectors with rank one. Such a measurement was considered below Eq. (2.4), there the measurement operators are

$E_\uparrow = |\uparrow\rangle\langle\uparrow|$ and $E_\downarrow = |\downarrow\rangle\langle\downarrow|$. In general, the measurement operators do not have to be projectors, they only need to satisfy the completeness Eq. (2.5).

For composite systems consisting of two subsystems, Alice and Bob, it is possible to perform *local measurements* on one of the subsystems. If a local measurement is done on Alice's subsystem, the subsystem of Bob remains unchanged. In this case, the measurement operators have the form $E_i = E_i^A \otimes \mathbb{1}^B$, with the identity operator $\mathbb{1}^B$ on Bob's Hilbert space. Similarly, measurement operators corresponding to local measurement on Bob's subsystem have the form $E_i = \mathbb{1}^A \otimes E_i^B$.

Finally, we also mention the concept of *quantum operations*, which is closely related to quantum measurements. Any set of measurement operators $\{E_i\}$ can also be called a quantum operation. The corresponding operators $E_i$ are then called *Kraus operators*. The action of a quantum operation $\{E_i\}$ on a density operator $\rho$ is given by

$$\Lambda(\rho) = \sum_i E_i \rho E_i^\dagger. \tag{2.9}$$

For composite systems, *local quantum operations* can be defined in the same way as it was done for local measurements. The importance of quantum operations lies in the fact that they describe the most general change of a quantum state possible in experiments. Quantum operations also play an important role in the study of noisy systems: noise is usually modeled as a quantum operation.

## 2.3 Reduced Density Operator

Sometimes one is only interested in one of the subsystems of a composite quantum system. This situation is captured by the concept of the *reduced density operator*. If the total system is described by the density operator $\rho^{AB}$, then the system of $A$ is described by the reduced density operator

$$\rho^A = \text{Tr}_B[\rho^{AB}], \tag{2.10}$$

where $\text{Tr}_B$ is called *partial trace* over the subsystem $B$. The partial trace is defined by

$$\text{Tr}_B[|a_1\rangle\langle a_2| \otimes |b_1\rangle\langle b_2|] = |a_1\rangle\langle a_2| \, \text{Tr}[|b_1\rangle\langle b_2|], \tag{2.11}$$

where $|a_1\rangle$ and $|a_2\rangle$ are any two vectors in $\mathcal{H}_A$, and $|b_1\rangle$ and $|b_2\rangle$ are any two vectors in $\mathcal{H}_B$. The trace on the right hand side is the usual trace for the subsystem $B$: $\text{Tr}[|b_1\rangle\langle b_2|] = \langle b_2|b_1\rangle$. In addition to Eq. (2.11), we also require that the partial trace is linear, i.e., $\text{Tr}_B[M^{AB} + N^{AB}] = \text{Tr}_B[M^{AB}] + \text{Tr}_B[N^{AB}]$ for any two operators $M^{AB}$ and $N^{AB}$. In this way, the partial trace is defined for all density operators. The physical meaning of the partial trace lies in the fact that it is the unique operation for obtaining correct measurement statistics for the subsystem $A$ [1, p. 105ff.].

## 2.4 Entropy and Mutual Information

The *von Neumann entropy* of a quantum state with density operator $\rho$ is defined as

$$S(\rho) = -\text{Tr}[\rho \log_2 \rho], \tag{2.12}$$

where the logarithm of the density operator $\rho$ is defined via its eigenvalues $\lambda_i$ and eigenstates $|i\rangle$ in the following way: $\log_2 \rho = \sum_i \log_2(\lambda_i) |i\rangle \langle i|$. With this definition, the entropy can be written as

$$S(\rho) = -\sum_i \lambda_i \log_2 \lambda_i, \tag{2.13}$$

where it is defined that $0 \log_2 0 = 0$.

The von Neumann entropy is the quantum version of the classical *Shannon entropy*. For a discrete random variable $X$ which can take a value $x$ with probability $p_x$, the Shannon entropy is defined as

$$H(X) = -\sum_x p_x \log_2 p_x. \tag{2.14}$$

Similar to the Shannon entropy, which measures the uncertainty of a classical random variable, the von Neumann entropy measures the uncertainty of a quantum state. Pure states represent full knowledge about a quantum system: their von Neumann entropy is zero. On the other hand, for a $d$-dimensional Hilbert space, maximal uncertainty is represented by the completely mixed density operator $\mathbb{1}/d$ with the von Neumann entropy $\log_2 d$.

For two parties, the von Neumann entropy can be used to define the *mutual information* between the parties. If the total state is given by the density operator $\rho^{AB}$ with reduced density operators $\rho^A$ and $\rho^B$, the mutual information is defined as

$$I(\rho^{AB}) = S(\rho^A) + S(\rho^B) - S(\rho^{AB}). \tag{2.15}$$

The mutual information is zero if the state is completely uncorrelated, i.e., if the density operator has the form $\rho^{AB} = \rho^A \otimes \rho^B$. Otherwise, the mutual information is greater than zero: it measures the amount of correlations between $A$ and $B$.

Closely related to the von Neumann entropy is the *quantum relative entropy*. For two density operators $\rho$ and $\sigma$ it is defined as

$$S(\rho||\sigma) = \text{Tr}[\rho \log_2 \rho] - \text{Tr}[\rho \log_2 \sigma] \tag{2.16}$$

if the support of $\rho$ is contained in the support of $\sigma$, and $S(\rho||\sigma) = +\infty$ otherwise. The quantum relative entropy is nonnegative, and zero if and only if $\rho = \sigma$. The mutual information defined in Eq. (2.15) can be written as the relative entropy between the density operator $\rho^{AB}$ and the tensor product of the reduced density operators $\rho^A \otimes \rho^B$ [2]:

$$I(\rho^{AB}) = S(\rho^{AB}||\rho^A \otimes \rho^B).$$ (2.17)

## 2.5 Distance Between Density Operators

Given two quantum states, how "close" are they to each other? This question, posed in [1, p. 403], can be answered by defining an appropriate distance onto the set of density operators. One important and frequently used distance is the *trace distance*

$$D_t(\rho, \sigma) = \frac{1}{2}\text{Tr}|\rho - \sigma|,$$ (2.18)

where $\rho$ and $\sigma$ are any two density operators, $\text{Tr}|M| = \text{Tr}\sqrt{M^\dagger M}$ is the trace norm of an operator $M$, and the square root of a Hermitian operator $M^\dagger M$ with nonnegative eigenvalues $\lambda_i$ and eigenstates $|i\rangle$ is defined as $\sqrt{M^\dagger M} = \sum_i \sqrt{\lambda_i}\,|i\rangle\,\langle i|$. The trace distance satisfies all properties of a general distance $D$:

- $D(\rho, \sigma) \geq 0$, and $D(\rho, \sigma) = 0$ holds if and only if $\rho = \sigma$,
- $D$ is symmetric: $D(\rho, \sigma) = D(\sigma, \rho)$,
- $D$ satisfies the triangle inequality: $D(\rho, \tau) \leq D(\rho, \sigma) + D(\sigma, \tau)$ for any three density operators $\rho$, $\sigma$, and $\tau$.

In quantum information theory, the trace distance has an important interpretation: $\frac{1}{2} + \frac{1}{2}D_t(\rho, \sigma)$ is the optimal probability of success for distinguishing two quantum states with density operators $\rho$ and $\sigma$ [3].

Another frequently used quantity is the *fidelity*. For two density operators $\rho$ and $\sigma$ it is defined as

$$F(\rho, \sigma) = \left(\text{Tr}\sqrt{\sqrt{\rho}\sigma\sqrt{\rho}}\right)^2.$$ (2.19)

The fidelity itself is not a distance, since it is one if and only if $\rho = \sigma$, and smaller than one otherwise. However, the fidelity can be used to define the *Bures distance*: $D_B(\rho, \sigma) = 2(1 - \sqrt{F(\rho, \sigma)})$, which satisfies all properties of a mathematical distance.

Both, the trace distance and the Bures distance have also another important property, namely they are *nonincreasing under quantum operations*:

$$D(\Lambda(\rho), \Lambda(\sigma)) \leq D(\rho, \sigma),$$ (2.20)

where $\rho$ and $\sigma$ are any two density operators, and $\Lambda$ is any quantum operation. This property is frequently used in quantum information theory, especially in studying entanglement and other quantum correlations.

Note that the inequality (2.20) does not follow from the general properties of a mathematical distance, and thus there exist distances which violate it. One such distance is the *Hilbert-Schmidt distance*

$$D_{HS}(\rho, \sigma) = \|\rho - \sigma\|, \tag{2.21}$$

where $\|M\| = \sqrt{\mathrm{Tr}[M^{\dagger}M]}$ is the Hilbert-Schmidt norm of an operator $M$. For the Hilbert-Schmidt distance violation of Eq. (2.20) was shown in [4, 5].

Finally, the relative entropy introduced in Eq. (2.16) is not a distance in the mathematical sense since it is not symmetric, and also does not satisfy the triangle inequality. However, the relative entropy is nonincreasing under quantum operations, i.e., it satisfies the inequality (2.20) [6].

# References

1. Nielsen, M.A., Chuang, I.L.: Quantum Computation and Quantum Information. Cambridge University Press, Cambridge (2000)
2. Vedral, V.: The role of relative entropy in quantum information theory. Rev. Mod. Phys. **74**, 197–234 (2002)
3. Fuchs, C., van de Graaf, J.: Cryptographic distinguishability measures for quantum-mechanical states. IEEE Trans. Inf. Theory **45**, 1216–1227 (1999)
4. Ozawa, M.: Entanglement measures and the Hilbert-Schmidt distance. Phys. Lett. A **268**, 158–160 (2000)
5. Piani, M.: Problem with geometric discord. Phys. Rev. A **86**, 034101 (2012)
6. Lindblad, G.: Completely positive maps and entropy inequalities. Commun. Math. Phys. **40**, 147–151 (1975)

# Chapter 3
# Quantum Entanglement

## 3.1 Definition

For two parties, Alice ($A$) and Bob ($B$), the state of the total quantum system can have product form[1]:

$$|\Psi\rangle = |a\rangle \otimes |b\rangle, \tag{3.1}$$

where the states $|a\rangle$ and $|b\rangle$ are elements of the corresponding local Hilbert spaces $\mathcal{H}_A$ and $\mathcal{H}_B$. States of the form given in Eq. (3.1) are not entangled, they are also called *separable*. However, not all states are separable, since quantum mechanics also allows superpositions which are not necessarily product:

$$|\Phi\rangle = \frac{1}{N}(|a_1\rangle \otimes |b_1\rangle + |a_2\rangle \otimes |b_2\rangle), \tag{3.2}$$

where $N$ assures normalization such that $\langle\Phi|\Phi\rangle = 1$. If $|\Phi\rangle$ cannot be written as a product, i.e., $|\Phi\rangle \neq |a\rangle \otimes |b\rangle$, the state is called *entangled*.

**Example** The singlet state $|\Phi\rangle = \frac{1}{\sqrt{2}}(|01\rangle - |10\rangle)$ is entangled, it cannot be written as a product.

A mixed state is separable if it can be written as a convex combination of pure product states [1]:

$$\rho_{\text{sep}} = \sum_i p_i |a_i\rangle \langle a_i| \otimes |b_i\rangle \langle b_i|. \tag{3.3}$$

The pure states $|a_i\rangle$ and $|b_i\rangle$ are elements of the local Hilbert spaces $\mathcal{H}_A$ and $\mathcal{H}_B$, and $p_i \geq 0$ are probabilities summing up to one: $\sum_i p_i = 1$. If the state cannot be written in this form, it is called entangled.

The idea behind this definition of entanglement is the following: suppose that Alice and Bob are able to produce any quantum state locally. In addition, they have access to a classical communication channel, such as a telephone. Then, Alice and

---

[1] Sometimes we write $|a\rangle |b\rangle$ or $|ab\rangle$ instead of $|a\rangle \otimes |b\rangle$.

© The Author(s) 2015
A. Streltsov, *Quantum Correlations Beyond Entanglement*,
SpringerBriefs in Physics, DOI 10.1007/978-3-319-09656-8_3

Bob can produce any separable state as given in Eq. (3.3) by the following procedure: Alice prepares the state $|a_i\rangle$ with the probability $p_i$, and lets Bob know which state she prepared. Depending on this information, Bob prepares the corresponding state $|b_i\rangle$. On the other hand, it is not possible to create entangled states such as the singlet state in this way.

## 3.2 Local Operations and Classical Communication

The process for creating separable states presented above belongs to the class of *local operations and classical communication* (LOCC), first introduced in [2]. This class of operations describes the most general procedure Alice and Bob can apply in quantum theory, if they are limited to classical communication only. The full mathematical description of these operations is demanding, and still subject of extensive research [3]. However, the general idea is simple, and will be explained in the following.

For two parties, Alice and Bob, a quantum operation $\Lambda_{LOCC}$ belongs to the class of LOCC, if it can be decomposed into the following steps:

1. One of the parties, e.g. Alice, performs a local measurement on her subsystem.
2. The outcome of the measurement is communicated *classically* to the other party, here Bob.
3. Depending on the received information, Bob performs a local measurement on his subsystem.
4. The outcome of Bob's measurement is communicated *classically* to Alice.
5. Depending on the received information, Alice performs a local measurement on her subsystem, and the process starts over at step 2.

The class of LOCC plays an important role in quantum information theory, especially when studying entanglement. As we have mentioned above, any separable state can be created with LOCC. On the other hand, LOCC cannot be used to create entangled states [4].

## 3.3 Entanglement as a Resource

Until the 1990s, quantum entanglement was mainly regarded as a physical curiosity: an exotic feature with no practical use. This situation started to change in 1991, when Ekert presented the first task in quantum information theory which was based on entanglement [5]. In his work, Ekert showed that if two parties, Alice and Bob, share a large amount of entangled singlet states, they can communicate in a completely secure way. This task is referred to as *quantum cryptography*, or *quantum key distribution*. This strong result should be compared to the classical cryptography as we use it today. The security of classical cryptography is mainly based on the conjecture that a large number is hard to factorize, whereas the quantum cryptography protocol presented by Ekert is provably secure.

Motivated by Ekert's result, several tasks involving entanglement have been presented in the following years. In 1992 Bennett and Wiesner showed that two entangled parties can communicate two classical bits by sending only one qubit, i.e., one quantum system on a two-dimensional Hilbert space [6]. This task is also known as *quantum dense coding*, since it suggests that two classical bits can be coded into one quantum bit.

Another application for entanglement has been proposed in [7]. The authors studied the task of communicating an unknown quantum state between two parties. An unknown quantum state cannot be communicated by classical means, which is a direct consequence of the fact that such a state cannot be cloned [8]. However, if the two parties share an entangled singlet, Bennett et al. showed that any unknown quantum bit can be perfectly communicated. This task is also known as *quantum teleportation*.

## 3.4 Entanglement Measures

The tasks presented above, namely quantum cryptography, dense coding and teleportation demonstrate the role of entanglement for a very special case. In particular, two parties, Alice and Bob, need to share entangled singlets in order to perform these tasks. However, a pure quantum state is not necessarily a singlet, and in a realistic scenario the quantum state is usually mixed. For this reason it is natural to ask whether a general mixed quantum state can also be used for some of these tasks.

The "usefulness" of a quantum state for one of the tasks presented above is usually quantified by the amount of entanglement contained in the state. One of the most popular quantifiers is the *distillable entanglement* [9]: it is defined as the maximal number of singlets that can be obtained per copy of a given mixed state via local operations and classical communication, if the number of copies goes to infinity.[2] The major disadvantage of the distillable entanglement is the fact that it is hard to evaluate. Thus, exact expressions are only known in a few special cases. For this reason, other quantifiers, known as *entanglement measures*, have been proposed in the literature. Any entanglement measure $E$ fulfills the following two properties [4]:

1. $E$ does not increase under local operations and classical communication,
2. $E$ vanishes on separable states.

For a *pure state* $|\psi\rangle^{AB}$ distributed between two parties, Alice and Bob, entanglement is usually quantified by the von Neumann entropy of the reduced density operator $\rho^A = \text{Tr}_B[|\psi\rangle\langle\psi|^{AB}]$:

---

[2] See also [4] for a formal definition.

$$E(|\psi\rangle^{AB}) = S(\rho^A) = -\sum_i \lambda_i \log_2 \lambda_i, \tag{3.4}$$

where $\lambda_i$ are the eigenvalues of $\rho^A$. The importance of this quantity in quantum information theory comes from the fact that it is equal to the distillable entanglement for all pure states [10].

For a *mixed state* $\rho^{AB}$, two main classes of entanglement measures are considered in the literature. These are

- convex roof measures and
- distance-based measures.

Any measure of entanglement $E$ which is defined on all pure states can be extended to mixed states via the following *convex roof* construction [11]:

$$E(\rho) = \inf_{\{p_i, |\psi_i\rangle\}} \sum_i p_i E(|\psi_i\rangle), \tag{3.5}$$

where the infimum is taken over all decompositions $\{p_i, |\psi_i\rangle\}$ of the given density operator $\rho$ with nonnegative probabilities $p_i$, i.e., $\rho = \sum_i p_i |\psi_i\rangle \langle\psi_i|$.

For bipartite systems, the *entanglement of formation* defined in [2] is one of the most popular and frequently used convex roof measures. For pure states it is defined as the von Neumann entropy of the reduced density operator in Eq. (3.4). The extension to mixed states is done via the convex roof construction in Eq. (3.5). Although the infimum in Eq. (3.5) is hard to evaluate in general, Wootters presented a closed expression for the entanglement of formation for all mixed states of two qubits [12]. For any such state, the entanglement of formation $E_f$ is given by

$$E_f(\rho) = h\left(\frac{1}{2} + \frac{1}{2}\sqrt{1 - C^2(\rho)}\right) \tag{3.6}$$

with the binary entropy $h(x) = -x\log_2 x - (1-x)\log_2(1-x)$, and the concurrence $C(\rho) = \max\{0, \lambda_1 - \lambda_2 - \lambda_3 - \lambda_4\}$, where $\lambda_i$ are the square roots of the eigenvalues of $\rho\tilde{\rho}$ in decreasing order, and $\tilde{\rho}$ is defined as $\tilde{\rho} = (\sigma_y \otimes \sigma_y)\rho^*(\sigma_y \otimes \sigma_y)$ with the Pauli matrix $\sigma_y = \begin{pmatrix} 0 & -i \\ i & 0 \end{pmatrix}$. The entanglement of formation satisfies the criteria for a proper entanglement measure given on page 13: it does not increase under local operations and classical communication and vanishes on separable states. While the second property is easy to verify, the first property was proven in [2].

The second main class of entanglement measures are measures based on distance proposed in [13]. All those measures can be written as

$$E(\rho) = \inf_{\sigma \in S} D(\rho, \sigma), \tag{3.7}$$

where $D$ is a distance, and the infimum is taken over the set of separable states $S$. If the distance $D$ does not increase under quantum operations, i.e.,

$$D(\Lambda(\rho), \Lambda(\sigma)) \leq D(\rho, \sigma) \tag{3.8}$$

for any quantum operation $\Lambda$ and any two states $\rho$ and $\sigma$, then the corresponding measure of entanglement does not increase under local operations and classical communication [13]. This property is satisfied by the relative entropy $S(\rho||\sigma) = \mathrm{Tr}[\rho \log_2 \rho] - \mathrm{Tr}[\rho \log_2 \sigma]$, although the relative entropy is not a distance in the mathematical sense. The corresponding measure of entanglement is called *relative entropy of entanglement*:

$$E_R(\rho) = \min_{\sigma \in \mathcal{S}} S(\rho||\sigma). \tag{3.9}$$

The relative entropy of entanglement is one of the most popular and widely studied measures of entanglement. One reason is the fact that the relative entropy itself plays an important role in quantum information theory [14]. Moreover, the relative entropy of entanglement is a powerful upper bound for the distillable entanglement [15].

We have already mentioned above that all distance-based entanglement measures do not increase under local operations and classical communication, if the distance satisfies Eq. (3.8). This is one of the properties any reasonable measure of entanglement should satisfy. Moreover, any entanglement measure should also vanish on separable states. This is also easily seen to be true for any distance $D(\rho, \sigma)$ which is zero if and only if $\rho = \sigma$, and larger than zero otherwise.

Finally, we mention the relation between three of the measures presented in this section, namely between the distillable entanglement $E_d$, the relative entropy of entanglement $E_R$, and the entanglement of formation $E_f$. As was shown in [15], these measures satisfy the inequality

$$E_d \leq E_R \leq E_f \tag{3.10}$$

for all mixed states, i.e., the relative entropy of entanglement is always between $E_d$ and $E_f$.

# References

1. Werner, R.F.: Quantum states with Einstein-Podolsky-Rosen correlations admitting a hidden-variable model. Phys. Rev. A **40**, 4277–4281 (1989)
2. Bennett, C.H., DiVincenzo, D.P., Smolin, J.A., Wootters, W.K.: Mixed-state entanglement and quantum error correction. Phys. Rev. A **54**, 3824–3851 (1996)
3. Chitambar, E., Leung, D., Mančinska, L., Ozols, M., Winter, A.: Everything you always wanted to know about LOCC (But Were Afraid to Ask). Commun. Math. Phys. **328**, 303–326 (2014)
4. Horodecki, R., Horodecki, P., Horodecki, M., Horodecki, K.: Quantum entanglement. Rev. Mod. Phys. **81**, 865–942 (2009)
5. Ekert, A.K.: Quantum cryptography based on Bell's theorem. Phys. Rev. Lett. **67**, 661–663 (1991)
6. Bennett, C.H., Wiesner, S.J.: Communication via one- and two-particle operators on Einstein-Podolsky-Rosen states. Phys. Rev. Lett. **69**, 2881–2884 (1992)

7. Bennett, C.H., et al.: Teleporting an unknown quantum state via dual classical and Einstein-Podolsky-Rosen channels. Phys. Rev. Lett. **70**, 1895–1899 (1993)
8. Wootters, W.K., Zurek, W.H.: A single quantum cannot be cloned. Nature (London) **299**, 802–803 (1982)
9. Bennett, C.H., et al.: Purification of noisy entanglement and faithful teleportation via noisy channels. Phys. Rev. Lett. **76**, 722–725 (1996)
10. Bennett, C.H., Bernstein, H.J., Popescu, S., Schumacher, B.: Concentrating partial entanglement by local operations. Phys. Rev. A **53**, 2046–2052 (1996)
11. Uhlmann, A.: Entropy and optimal decompositions of states relative to a maximal commutative subalgebra. Open Syst. Inf. Dyn. **5**, 209–228 (1998)
12. Wootters, W.K.: Entanglement of formation of an arbitrary state of two qubits. Phys. Rev. Lett. **80**, 2245–2248 (1998)
13. Vedral, V., Plenio, M.B., Rippin, M.A., Knight, P.L.: Quantifying entanglement. Phys. Rev. Lett. **78**, 2275–2279 (1997)
14. Vedral, V.: The role of relative entropy in quantum information theory. Rev. Mod. Phys. **74**, 197–234 (2002)
15. Horodecki, M., Horodecki, P., Horodecki, R.: Limits for entanglement measures. Phys. Rev. Lett. **84**, 2014–2017 (2000)

# Chapter 4
# Quantum Correlations Beyond Entanglement

## 4.1 Definition

A mixed state shared by two parties, Alice and Bob, is called *classically correlated* if it can be written as [1]

$$\rho_{cc} = \sum_{i,j} p_{ij} |i\rangle\langle i|^A \otimes |j\rangle\langle j|^B, \tag{4.1}$$

where $\{|i\rangle^A\}$ are orthogonal states on Alice's Hilbert space $\mathcal{H}_A$ and $\{|j\rangle^B\}$ are orthogonal states on Bob's Hilbert space $\mathcal{H}_B$. The probabilities $p_{ij}$ are nonnegative and sum up to one: $\sum_{i,j} p_{ij} = 1$. Otherwise the state is called *quantum correlated*. Note that every classically correlated state is also separable. On the other hand, a separable state $\rho_{sep} = \sum_i p_i |a_i\rangle\langle a_i| \otimes |b_i\rangle\langle b_i|$ is not necessarily classically correlated, since the states $\{|a_i\rangle\}$ and $\{|b_i\rangle\}$ do not have to be orthogonal. Moreover, a pure state is quantum correlated if and only if the state is entangled, i.e., both concepts are equivalent for pure states. For this reason, we will discuss mixed states in the following.

The intuition behind this definition of classically correlated states comes from the fact that these states are not disturbed by certain local von Neumann measurements on Alice's and Bob's subspaces. The measurement operators corresponding to these non-disturbing von Neumann measurements are given by $E_i^A = |i\rangle\langle i|^A$ and $E_j^B = |j\rangle\langle j|^B$. In a similar way we can also define a class of quantum states which is not disturbed under certain von Neumann measurements on the subspace of one party (e.g. Alice) only. In this case the state has the form

$$\rho_{cq} = \sum_i p_i |i\rangle\langle i|^A \otimes \rho_i^B, \tag{4.2}$$

where $|i\rangle^A$ are orthogonal states on Alice's Hilbert space $\mathcal{H}_A$, $\rho_i^B$ are states on Bob's Hilbert space $\mathcal{H}_B$, and the nonnegative probabilities $p_i$ sum up to one. These states are called *classical-quantum* states [2, 3]. The corresponding von Neumann measurement on Alice's subsystem which does not disturb the total state is given by the measurement operators $E_i^A = |i\rangle\langle i|^A$. Similarly, a *quantum-classical* state

© The Author(s) 2015

A. Streltsov, *Quantum Correlations Beyond Entanglement*,
SpringerBriefs in Physics, DOI 10.1007/978-3-319-09656-8_4

has the form $\rho_{qc} = \sum_i p_i \rho_i^A \otimes |i\rangle \langle i|^B$. Such a state is not disturbed by a local von Neumann measurement on Bob's subspace with measurement operators $E_i^B = |i\rangle \langle i|^B$.

## 4.2 Measures of Quantum Correlations

A measure of entanglement can be defined via the usefulness of a quantum state to perform certain tasks. The figure of merit is the distillable entanglement, which quantifies how many singlets can be extracted per copy of a given quantum state via local operations and classical communication, if many copies of the same state are available. Since singlets can be used for many tasks in quantum information theory, e.g., quantum cryptography, dense coding and teleportation, the distillable entanglement is directly related to the performance of these tasks.

For general quantum correlations the situation is less clear, since the definition of "distillable quantum correlations" is meaningless, at least if the concept of local operations and classical communication is considered. The reason for this is the fact that local operations and classical communication can be used to create an arbitrary amount of quantum correlations [4, 5]. This means that a measure of "distillable quantum correlations" would be infinite for all quantum states. However, several other approaches to quantify quantum correlations have been proposed in the literature. The most important measures of quantum correlations will be presented in the following.

### 4.2.1 Quantum Discord

Quantum discord is historically the first measure of quantum correlations beyond entanglement [6–8]. The definition of quantum discord is based on the fact that in classical information theory the mutual information between two random variables $X$ and $Y$ can be expressed in two different ways, namely

$$I(X : Y) = H(X) + H(Y) - H(X, Y),$$
$$J(X : Y) = H(X) - H(X|Y). \tag{4.3}$$

Here, $H(X) = -\sum_x p_x \log_2 p_x$ is the classical Shannon entropy of the random variable $X$, where $p_x$ is the probability that the random variable $X$ takes the value $x$. $H(X, Y)$ is the joint entropy of both variables $X$ and $Y$. The conditional entropy $H(X|Y)$ is defined as

$$H(X|Y) = \sum_y p_y H(X|y), \tag{4.4}$$

where $p_y$ is the probability that the random variable $Y$ takes the value $y$, and $H(X|y)$ is the entropy of the variable $X$ conditioned on the variable $Y$ taking the value $y$: $H(X|y) = -\sum_x p_{x|y} \log_2 p_{x|y}$, and $p_{x|y}$ is the probability of $x$ given $y$.

The equality of $I$ and $J$ for classical random variables follows from Bayes' rule $p_{x|y} = p_{xy}/p_y$, which can be used to show that $H(X|Y) = H(X,Y) - H(Y)$. However, as was noticed in [7], $I$ and $J$ are no longer equal if quantum theory is applied. In particular, for a quantum state $\rho^{AB}$ the mutual information between $A$ and $B$ is given by

$$I(\rho^{AB}) = S(\rho^A) + S(\rho^B) - S(\rho^{AB}) \tag{4.5}$$

with the von Neumann entropy $S$, and the reduced density operators $\rho^A = \mathrm{Tr}_B[\rho^{AB}]$ and $\rho^B = \mathrm{Tr}_A[\rho^{AB}]$. This expression is the generalization of the classical mutual information $I(X:Y)$ to the quantum theory.

On the other hand, the generalization of $J(X:Y)$ is not completely straightforward. Ollivier and Zurek have proposed the following way to generalize $J$ to the quantum theory [7]: for a bipartite quantum state $\rho^{AB}$ they defined the conditional entropy of $A$ conditioned on a measurement on $B$:

$$S(A|\{\Pi_i^B\}) = \sum_i p_i S(\rho_i^A), \tag{4.6}$$

where $\{\Pi_i^B\}$ are measurement operators corresponding to a von Neumann measurement on the subsystem $B$, i.e., orthogonal projectors with rank one. The probability $p_i$ for obtaining the outcome $i$ is given by $p_i = \mathrm{Tr}[\Pi_i^B \rho^{AB}]$, and the corresponding post-measurement state of the subsystem $A$ is given by $\rho_i^A = \mathrm{Tr}_B[\Pi_i^B \rho^{AB}]/p_i$. The quantity $J$ can now be extended to quantum states as follows [7]:

$$J(\rho^{AB})_{\{\Pi_i^B\}} = S(\rho^A) - S(A|\{\Pi_i^B\}), \tag{4.7}$$

where the index $\{\Pi_i^B\}$ clarifies that the value depends on the choice of the measurement operators $\Pi_i^B$. The quantity $J$ represents the amount of information gained about the subsystem $A$ by measuring the subsystem $B$ [7].

*Quantum discord* is the difference of these two inequivalent expressions for the mutual information, minimized over all von Neumann measurements:

$$\delta^{B|A}(\rho^{AB}) = \min_{\{\Pi_i^B\}} \left[ I(\rho^{AB}) - J(\rho^{AB})_{\{\Pi_i^B\}} \right], \tag{4.8}$$

where the minimum over all von Neumann measurements is taken in order to have a measurement-independent expression [7]. As was also shown in [7], quantum discord is nonnegative, and is equal to zero on quantum-classical states only. These are states of the form $\rho_{qc} = \sum_i p_i \rho_i^A \otimes |i\rangle \langle i|^B$.

A closely related quantity was proposed by Henderson and Vedral in [8]. The authors aimed to quantify *classical correlations* in quantum states by defining a measure of classical correlations $C_B$ which is equal to $J$ given in Eq. (4.7), maximized over all positive operator-valued measures (POVMs) on the subsystem $B$:

$$C_B(\rho^{AB}) = \sup_{\{M_i^B\}} J(\rho^{AB})_{\{M_i^B\}}. \tag{4.9}$$

Here, $M_i^B$ are POVM elements on the subsystem $B$, and $J(\rho^{AB})_{\{M_i^B\}}$ is the generalization of Eq. (4.7) to POVMs:

$$J(\rho^{AB})_{\{M_i^B\}} = S(\rho^A) - S(A|\{M_i^B\}) \tag{4.10}$$

with $S(A|\{M_i^B\}) = \sum_i p_i S(\rho_i^A)$. The measurement probabilities are now given by $p_i = \mathrm{Tr}[M_i^B \rho^{AB}]$, and the corresponding post-measurement state of the subsystem $A$ is given by $\rho_i^A = \mathrm{Tr}_B[M_i^B \rho^{AB}]/p_i$.

In today's literature, quantum discord is frequently defined as the difference between the mutual information $I$, and the amount of classical correlations $C_B$ [9]:

$$D^{B|A}(\rho^{AB}) = I(\rho^{AB}) - C_B(\rho^{AB}). \tag{4.11}$$

This measure is in general different from the original quantum discord $\delta^{B|A}$ proposed by Ollivier and Zurek. However, this quantity is also nonnegative, and vanishes on quantum-classical states only [10]. Quantum discord as defined in Eq. (4.11) is related to the entanglement of formation $E_f$ via the Koashi-Winter relation [11, 12]:

$$D^{B|A}(\rho^{AB}) = E_f(\rho^{AC}) - S(\rho^{AB}) + S(\rho^B), \tag{4.12}$$

where the total state $\rho^{ABC}$ is pure, i.e., $\rho^{ABC} = |\psi\rangle\langle\psi|^{ABC}$.

### 4.2.2 General Measures of Quantum Correlations

Postulates for general measures of quantum correlations have been proposed in [13]. There the authors identify three necessary conditions every measure of quantum correlations $Q$ should satisfy. These conditions are:

1. $Q$ is nonnegative,
2. $Q$ is invariant under local unitary operations,
3. $Q$ is zero on classically correlated states.

Note that both versions of quantum discord, $\delta$ and $D$, satisfy all these criteria. In the following we will present main measures of general quantum correlations apart from quantum discord.

*Information deficit* is a measure of quantum correlations which was originally based on the task of extracting work from a heat bath using a quantum state [1, 2]. In particular, the amount of extractable work from a heat bath of temperature $T$ using a mixed state $\rho$ of $n$ qubits is given by

$$W = kT\{n - S(\rho)\}, \tag{4.13}$$

where $k$ is the Boltzmann constant and $S$ is the von Neumann entropy. However, if the state is shared by two parties, Alice and Bob, each of them having access to the local subsystem only, the amount of extractable work $W'$ will in general be different from $W$. If Alice is allowed to perform a single von Neumann measurement on her local system and send the resulting state to Bob, the maximal amount of work which Bob can extract from the resulting state in this way is given by

$$W' = W - kT \cdot \Delta^{A|B}(\rho^{AB}), \tag{4.14}$$

where $\Delta^{A|B}$ is known as the *one-way information deficit* [2]:

$$\Delta^{A|B}(\rho^{AB}) = \min_{\{\Pi_i^A\}} S(\rho^{AB} || \sum_i \Pi_i^A \rho^{AB} \Pi_i^A). \tag{4.15}$$

$S(\rho||\sigma)$ is the relative entropy between the states $\rho$ and $\sigma$, and the minimum is taken over local von Neumann measurements $\{\Pi_i^A\}$ on the subsystem $A$. The one-way information deficit is zero on classical-quantum states only, and can also be written as the minimal relative entropy between the given state $\rho^{AB}$ and the set of classical-quantum states $CQ$ [14]:

$$\Delta^{A|B}(\rho^{AB}) = \min_{\sigma^{AB} \in CQ} S(\rho^{AB} || \sigma^{AB}). \tag{4.16}$$

For this reason, this quantity is also called *relative entropy of discord*. In a similar way, it is possible to define the *relative entropy of quantumness* as the minimal relative entropy between $\rho^{AB}$ and the set of classically correlated states $CC$ [15]:

$$Q_R(\rho^{AB}) = \min_{\sigma^{AB} \in CC} S(\rho^{AB} || \sigma^{AB}). \tag{4.17}$$

Inspired by the expression for the relative entropy of discord as the minimal relative entropy between a given state and the set of classical-quantum states $CQ$, Dakić et al. defined the *geometric measure of discord* as the minimal squared Hilbert-Schmidt distance between a given state $\rho^{AB}$ and $CQ$ [4]:

$$D_G^{A|B}(\rho^{AB}) = \min_{\sigma^{AB} \in CQ} \left\| \rho^{AB} - \sigma^{AB} \right\|^2 \tag{4.18}$$

with the Hilbert-Schmidt norm $\|M\| = \sqrt{\text{Tr}[M^\dagger M]}$. The main advantage of the geometric measure of discord was already presented in the original work by Dakić et al.: this measure has an analytical expression for all two-qubit states [4]. If $\rho^{AB}$ is a two-qubit state, then the geometric measure of discord can be written as [4]

$$D_G^{A|B}(\rho^{AB}) = \frac{1}{4}(a^2 + \text{Tr}[E^T E] - k_{\max}),  \tag{4.19}$$

where $a$ is a 3-dimensional vector with entries $a_i = \text{Tr}[(\sigma_i \otimes \mathbb{1})\rho^{AB}]$, and $E$ is the $3 \times 3$ correlation tensor with components $E_{ij} = \text{Tr}[(\sigma_i \otimes \sigma_j)\rho^{AB}]$. The Pauli operators $\sigma_i$ are given as $\sigma_1 = \begin{pmatrix} 0 & 1 \\ 1 & 0 \end{pmatrix}$, $\sigma_2 = \begin{pmatrix} 0 & -i \\ i & 0 \end{pmatrix}$, and $\sigma_3 = \begin{pmatrix} 1 & 0 \\ 0 & -1 \end{pmatrix}$. Finally, $k_{\max}$ is the largest eigenvalue of the real matrix $aa^T + EE^T$.

# References

1. Oppenheim, J., Horodecki, M., Horodecki, P., Horodecki, R.: Thermodynamical approach to quantifying quantum correlations. Phys. Rev. Lett. **89**, 180402 (2002)
2. Horodecki, M., et al.: Local versus nonlocal information in quantum-information theory: formalism and phenomena. Phys. Rev. A **71**, 062307 (2005)
3. Piani, M., Horodecki, P., Horodecki, R.: No-local-broadcasting theorem for multipartite quantum correlations. Phys. Rev. Lett. **100**, 090502 (2008)
4. Dakić, B., Vedral, V., Brukner, Č.: Necessary and sufficient condition for nonzero quantum discord. Phys. Rev. Lett. **105**, 190502 (2010)
5. Streltsov, A., Kampermann, H., Bruß, D.: Behavior of quantum correlations under local noise. Phys. Rev. Lett. **107**, 170502 (2011)
6. Zurek, W.H.: Einselection and decoherence from an information theory perspective. Ann. Phys. (Leipzig) **9**, 855–864 (2000)
7. Ollivier, H., Zurek, W.H.: Quantum discord: a measure of the quantumness of correlations. Phys. Rev. Lett. **88**, 017901 (2001)
8. Henderson, L., Vedral, V.: Classical, quantum and total correlations. J. Phys. A: Math. Gen. **34**, 6899 (2001)
9. Datta, A.: Studies on the Role of Entanglement in Mixed-state Quantum Computation. Ph.D. thesis, University of New Mexico (2008). arXiv:0807.4490v1
10. Datta, A.: A Condition for the Nullity of Quantum Discord (2010). arXiv:1003.5256v2
11. Koashi, M., Winter, A.: Monogamy of quantum entanglement and other correlations. Phys. Rev. A **69**, 022309 (2004)
12. Fanchini, F.F., Cornelio, M.F., de Oliveira, M.C., Caldeira, A.O.: Conservation law for distributed entanglement of formation and quantum discord. Phys. Rev. A **84**, 012313 (2011)
13. Brodutch, A., Modi, K.: Criteria for measures of quantum correlations. Quantum Inf. Comput. **12**, 0721–0742 (2012)
14. Modi, K., Paterek, T., Son, W., Vedral, V., Williamson, M.: Unified view of quantum and classical correlations. Phys. Rev. Lett. **104**, 080501 (2010)
15. Piani, M., et al.: All nonclassical correlations can be activated into distillable entanglement. Phys. Rev. Lett. **106**, 220403 (2011)

# Chapter 5
# Quantum Discord in Quantum Information Theory

## 5.1 Remote State Preparation

### 5.1.1 Deterministic Remote State Preparation

The role of quantum discord in the task of *remote state preparation* was considered by Dakić et al. [1]. In this task Alice aims to remotely prepare Bob's system in the quantum state

$$|\psi\rangle = \frac{1}{\sqrt{2}}(|0\rangle + e^{i\phi}|1\rangle). \tag{5.1}$$

To this end Alice and Bob have access to an additional shared quantum state and a classical communication channel. In contrast to the standard quantum teleportation [2], which can be applied to remotely prepare an arbitrary quantum state by making use of a shared singlet and two bits of classical communication, remote preparation of the state given in Eq. (5.1) requires a shared singlet supported by only one classical bit [3]. To achieve this task, Alice applies a von Neumann measurement in the basis $\{|\psi_\perp\rangle\langle\psi_\perp|, |\psi\rangle\langle\psi|\}$ on her part of the singlet, where the state $|\psi_\perp\rangle = (|0\rangle - e^{i\phi}|1\rangle)/\sqrt{2}$ is orthogonal to $|\psi\rangle$. Depending on her outcome, Bob's system is found in one of the states $|\psi\rangle\langle\psi|$ or $|\psi_\perp\rangle\langle\psi_\perp|$. By sending the outcome of her measurement to Bob—which implies sending one classical bit—he either finds his system in the desired state $|\psi\rangle\langle\psi|$, or can correct his state $|\psi_\perp\rangle\langle\psi_\perp|$ by applying the Pauli operator $\sigma_z$.

Note that the Bloch vector of the state $|\psi\rangle$ in Eq. (5.1) lies in the equatorial plane of the Bloch sphere orthogonal to the z axis. Moreover, the $\sigma_z$ operation which is applied by Bob to the state $|\psi_\perp\rangle\langle\psi_\perp|$ can be regarded as a $\pi$ rotation around the z axis of the corresponding Bloch vector. In a similar way Alice can remotely prepare any pure state in a fixed equatorial plane of the Bloch sphere. If $s$ is the Bloch vector of the state $|s\rangle\langle s|$ Alice wishes to prepare and $\beta$ is a normalized vector orthogonal to the corresponding equatorial plane, Alice can achieve this task by performing a von Neumann measurement in the basis $\{|-s\rangle\langle-s|, |s\rangle\langle s|\}$ on her part of the singlet and send the outcome to Bob. Depending on the measurement outcome, Bob either finds

© The Author(s) 2015
A. Streltsov, *Quantum Correlations Beyond Entanglement*,
SpringerBriefs in Physics, DOI 10.1007/978-3-319-09656-8_5

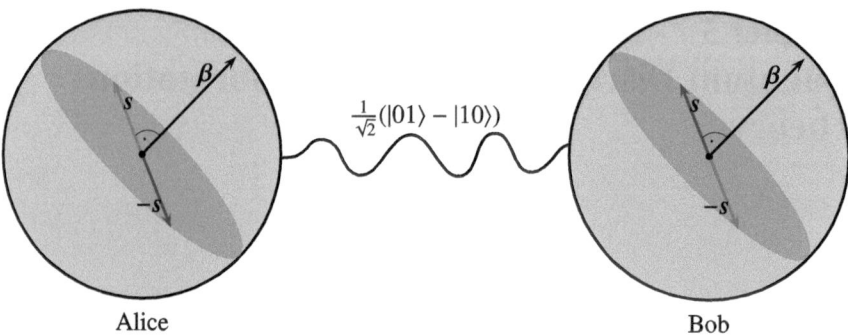

**Fig. 5.1** Remote state preparation. Alice can remotely prepare any state $|s\rangle \langle s|$ with Bloch vector $s$ on a fixed equatorial plane of the Bloch sphere by performing a von Neumann measurement in the basis $\{|-s\rangle \langle -s|, |s\rangle \langle s|\}$ on her part of the singlet and by sending the outcome of the measurement to Bob. Depending on the outcome, Bob's system is found in one of the states $|s\rangle \langle s|$ or $|-s\rangle \langle -s|$. Bob can correct the latter by applying a $\pi$ rotation around the direction $\beta$ orthogonal to the corresponding equatorial plane

his system in the state $|s\rangle \langle s|$ or $|-s\rangle \langle -s|$ and can correct the latter by applying a $\pi$ rotation around the direction $\beta$, see also Fig. 5.1 for illustration.

### 5.1.2  Remote State Preparation in the Presence of Noise

So far we considered deterministic remote state preparation where Alice could remotely prepare the desired state with certainty. However, this is not possible in general if the state shared by Alice and Bob is mixed, and the above procedure will leave Bob's system in a mixed state with Bloch vector $r$. The aim of Alice in this case is to adjust her measurement such that Bob's final Bloch vector $r$ becomes as close as possible to the desired vector $s$. As a quantifier of the performance of this procedure Dakić et al. introduced the payoff-function [1]

$$\mathcal{P} = (r \cdot s)^2. \tag{5.2}$$

For a pure state $|s\rangle$ with Bloch vector $s$ and a mixed state $\rho$ with Bloch vector $r$ the payoff function is directly related to the fidelity $\langle s|\rho|s\rangle$ between the two states. This can be seen by writing the fidelity explicitly as $\langle s|\rho|s\rangle = (1 + r \cdot s)/2$, and thus we get the desired relation $\mathcal{P} = (2\langle s|\rho|s\rangle - 1)^2$.

The aim of Alice is to maximize the payoff function for a given Bloch vector $s$. As was shown in [1], this maximization can be performed for any mixed two-qubit state shared by Alice and Bob. Note that any such state admits the following representation:

$$\rho = \frac{1}{4}\left(\mathbb{1} \otimes \mathbb{1} + \sum_{i=1}^{3} a_i \cdot \sigma_i \otimes \mathbb{1} + \sum_{j=1}^{3} b_j \cdot \mathbb{1} \otimes \sigma_j + \sum_{k,l=1}^{3} E_{kl} \cdot \sigma_k \otimes \sigma_l\right). \quad (5.3)$$

The vectors $a = (a_1, a_2, a_3)$ and $b = (b_1, b_2, b_3)$ are the Bloch vectors of Alice's and Bob's local state respectively, and $E_{kl} = \mathrm{Tr}[\sigma_k \otimes \sigma_l \rho]$ are the elements of the correlation tensor $E$. If Alice applies a von Neumann measurement in the basis $\{|\alpha\rangle\langle\alpha|, |-\alpha\rangle\langle-\alpha|\}$ on her part of the mixed state, she obtains one of two outcomes with corresponding probabilities $p_\alpha$ and $p_{-\alpha} = 1 - p_\alpha$. By using the relation $|\alpha\rangle\langle\alpha| = \frac{1}{2}(\mathbb{1} + \alpha \cdot \sigma)$, where $\sigma = (\sigma_1, \sigma_2, \sigma_3)$ is a vector containing the Pauli matrices, we can express the probability $p_\alpha$ as follows:

$$p_\alpha = \mathrm{Tr}\left[|\alpha\rangle\langle\alpha| \otimes \mathbb{1}\rho\right] = \frac{1}{2}(1 + \alpha \cdot a). \quad (5.4)$$

Conditioned on the outcome of this measurement, the state of Bob's system is found to be

$$\rho_\alpha^B = \frac{\mathrm{Tr}_A\left[|\alpha\rangle\langle\alpha| \otimes \mathbb{1}\rho\right]}{p_\alpha}. \quad (5.5)$$

Recall that this state can also be written in the form $\rho_\alpha^B = (\mathbb{1} + b_\alpha \cdot \sigma)/2$ with the Bloch vector $b_\alpha$. By inserting Eqs. (5.3) and (5.4) into Eq. (5.5) the Bloch vector $b_\alpha$ can be written explicitly as

$$b_\alpha = \frac{b + E^T \alpha}{1 + \alpha \cdot a}, \quad (5.6)$$

where $E^T$ is the transposed correlation tensor $E$.

In the next steps Alice and Bob follow the same procedure as for the deterministic remote state preparation discussed above (see Fig. 5.1). In particular, Alice sends the outcome of her measurement to Bob, who applies a $\pi$ rotation $R_\pi$ around the direction $\beta$ conditioned on the outcome of the measurement. After these steps the Bloch vector $r_\alpha$ of Bob's final state takes the form

$$r_\alpha = p_\alpha b_\alpha + p_{-\alpha} R_\pi b_{-\alpha}. \quad (5.7)$$

Note that this procedure is optimal if the state shared by Alice and Bob is a singlet, and if Alice chooses her measurement basis $\{|\alpha\rangle\langle\alpha|, |-\alpha\rangle\langle-\alpha|\}$ such that $\alpha = -s$, where $s$ is the Bloch vector of the state $|s\rangle$ Alice wishes to prepare. In this case the Bloch vector of Bob's final state $r_{-s}$ is equal to $s$.

For evaluating the payoff function $\mathcal{P} = (r \cdot s)^2$ it is crucial to note that any vector $x$ satisfies the following equality: $(R_\pi x) \cdot s = -x \cdot s$. This equality can be proven by using the invariance of the scalar product under rotations, i.e., $(R_\pi x) \cdot (R_\pi y) = x \cdot y$ for any two vectors $x$ and $y$. With this in mind, we see that $(R_\pi x) \cdot s = (R_\pi^2 x) \cdot (R_\pi s) = -x \cdot s$, where in the last step we used the fact that double application of the rotation $R_\pi$ does not change the vector, i.e., $R_\pi^2 x = x$, and that the rotation $R_\pi$ applied to the vector $s$ takes it to $-s$, see Fig. 5.1. Using these results the product $r \cdot s$ takes the following form: $r \cdot s = p_\alpha b_\alpha \cdot s - p_{-\alpha} b_{-\alpha} \cdot s$. By using Eqs. (5.4) and (5.6) this product can also be written as $r \cdot s = \alpha E s$, and the payoff function reduces to

$$\mathcal{P} = (\alpha E s)^2. \tag{5.8}$$

This expression is valid for any von Neumann measurement of Alice in the basis $\{|\alpha\rangle \langle \alpha|, |-\alpha\rangle \langle -\alpha|\}$. Maximal payoff is achieved if the vector $\alpha$ is parallel to the vector $E s$, i.e., $\alpha = E s / \sqrt{(E s)^2}$, and the maximum is thus given by the simple expression

$$\mathcal{P}_{\max} = (E s)^2. \tag{5.9}$$

### 5.1.3 Average Payoff

Following the discussion in [1], we will assess the average quality of the remote preparation procedure by the average payoff $\langle \mathcal{P}_{\max} \rangle$, where the mean value is taken over all Bloch vectors $s$ for a fixed direction $\beta$ (see Fig. 5.1). The calculation of $\langle \mathcal{P}_{\max} \rangle$ can be simplified by introducing a rotation matrix $R$ which rotates the vector $\beta$ onto the z axis: $\tilde{\beta} = R\beta = (0, 0, 1)$, and an arbitrary vector $x$ is rotated to $\tilde{x} = Rx$. If we further introduce the rotated correlation tensor $\tilde{E} = RER^T$, we see that rotations do not change the maximal payoff:

$$\mathcal{P}_{\max} = (E s)^2 = \left( \tilde{E} \tilde{s} \right)^2. \tag{5.10}$$

Since the vectors $s$ and $\beta$ are orthogonal, the same is also true for $\tilde{s}$ and $\tilde{\beta} = (0, 0, 1)$, and thus the normalized vector $\tilde{s}$ takes the form $\tilde{s} = (\cos \phi, \sin \phi, 0)$. Using these tools we are now in position to give a closed expression for the average payoff:

$$\langle \mathcal{P}_{\max} \rangle = \frac{1}{2\pi} \int_0^{2\pi} d\phi \left( \tilde{E} \tilde{s} \right)^2 = \frac{1}{2} \text{Tr} \left[ E^T E \right] - \frac{1}{2} (E\beta)^2. \tag{5.11}$$

This expression can be proven by writing $\left( \tilde{E} \tilde{s} \right)^2$ explicitly as $\left( \tilde{E} \tilde{s} \right)^2 = \sum_{i=1}^3$ $(\tilde{E}_{i1} \cos \phi + \tilde{E}_{i2} \sin \phi)^2$ and evaluating the integral in Eq. (5.11): $\frac{1}{2\pi} \int_0^{2\pi} d\phi \left( \tilde{E} \tilde{s} \right)^2 = \frac{1}{2} \sum_{i=1}^3 (\tilde{E}_{i1}^2 + \tilde{E}_{i2}^2)$. By using the relations $\text{Tr}[\tilde{E}^T \tilde{E}] = \sum_{i,j=1}^3 \tilde{E}_{ij}^2$ and $\tilde{E}\tilde{\beta} =$

$(\tilde{E}_{13}, \tilde{E}_{23}, \tilde{E}_{33})$ the integral can further be expressed as $\frac{1}{2\pi} \int_0^{2\pi} d\phi \left(\tilde{E}\tilde{s}\right)^2 = \frac{1}{2}\text{Tr}[\tilde{E}^T \tilde{E}] - \frac{1}{2}\left(\tilde{E}\tilde{\boldsymbol{\beta}}\right)^2$. The desired equality (5.11) follows by noting that both terms $\text{Tr}[\tilde{E}^T \tilde{E}]$ and $\left(\tilde{E}\tilde{\boldsymbol{\beta}}\right)^2$ are invariant under rotations, i.e., $\text{Tr}[\tilde{E}^T \tilde{E}] = \text{Tr}[E^T E]$ and $\left(\tilde{E}\tilde{\boldsymbol{\beta}}\right)^2 = (E\boldsymbol{\beta})^2$.

### 5.1.4 The Role of Quantum Correlations

Note that the matrix $E^T E$ has three nonnegative eigenvalues $\lambda_1 \geq \lambda_2 \geq \lambda_3$, and the average payoff $\langle \mathcal{P}_{\max} \rangle$ satisfies the inequality

$$\langle \mathcal{P}_{\max} \rangle \geq \min_{\boldsymbol{\beta}} \langle \mathcal{P}_{\max} \rangle = \frac{1}{2}(\lambda_2 + \lambda_3), \tag{5.12}$$

where the minimum is taken over all normalized vectors $\boldsymbol{\beta}$. This can be seen by noting that $\min_{\boldsymbol{\beta}} \langle \mathcal{P}_{\max} \rangle = \frac{1}{2}\text{Tr}\left[E^T E\right] - \frac{1}{2} \max_{\boldsymbol{\beta}} (E\boldsymbol{\beta})^2$, and the latter maximization can be performed as $\max_{\boldsymbol{\beta}} (E\boldsymbol{\beta})^2 = \max_{\boldsymbol{\beta}}(\boldsymbol{\beta} E^T E\boldsymbol{\beta}) = \lambda_1$ (see p. 176 in [4]). The result in Eq. (5.12) is obtained by noting that $\text{Tr}\left[E^T E\right] = \sum_{i=1}^{3} \lambda_i$.

According to [1], the quantity $\min_{\boldsymbol{\beta}} \langle \mathcal{P}_{\max} \rangle$ can be regarded as the quantifier of efficiency for remote state preparation in the worst case, i.e., for the most inconvenient choice of the direction $\boldsymbol{\beta}$. The relation to quantum discord is established by noting that in a large number of scenarios this expression corresponds to the geometric measure of discord of the shared state. In general, the geometric measure of discord is defined as [5]

$$D_G(\rho) = \min_{\sigma \in CQ} ||\rho - \sigma||^2, \tag{5.13}$$

where the minimum is taken over all classical-quantum states $\sigma$, and $||M|| = \sqrt{\text{Tr}[M^\dagger M]}$ is the Hilbert-Schmidt norm of the operator $M$. For states of two qubits as given in Eq. (5.3) the geometric measure of discord takes the form [5]

$$D_G(\rho) = \frac{1}{4}(a^2 + \text{Tr}[E^T E] - k_{\max}), \tag{5.14}$$

where $\boldsymbol{a}$ is the Bloch vector of Alice's subsystem, and $k_{\max}$ is the largest eigenvalue of the matrix $\boldsymbol{a}\boldsymbol{a}^T + EE^T$.

By the singular value decomposition of the correlation tensor $E$ it follows that the eigenvalues $\lambda_1 \geq \lambda_2 \geq \lambda_3$ of the matrix $E^T E$ are also eigenvalues of $EE^T$. Let now $\boldsymbol{\lambda}_1$ be the normalized eigenvector of $EE^T$ corresponding to the largest eigenvalue $\lambda_1$ and consider the situation where the Bloch vector $\boldsymbol{a}$ of Alice's subsystem is parallel to $\boldsymbol{\lambda}_1$, i.e., $\boldsymbol{a} = \sqrt{a^2}\boldsymbol{\lambda}_1$. In this case the eigenvalues of the matrix $\boldsymbol{a}\boldsymbol{a}^T + EE^T$ are given as $\{a^2 + \lambda_1, \lambda_2, \lambda_3\}$, and the largest eigenvalue becomes $k_{\max} = a^2 + \lambda_1$. Inserting this result into Eq. (5.14) and recalling that $\text{Tr}[E^T E] = \sum_{i=1}^{3} \lambda_i$ we obtain the expression $D_G(\rho) = \frac{1}{4}(\lambda_2 + \lambda_3)$. Together with Eq. (5.12) this result implies that for the particular family of shared states $\rho$ where the Bloch vector $\boldsymbol{a}$ is parallel to $\lambda_1$ the average payoff is bounded below by the geometric measure of discord:

$$\langle \mathcal{P}_{\max} \rangle \geq \min_{\boldsymbol{\beta}} \langle \mathcal{P}_{\max} \rangle = 2D_G(\rho). \tag{5.15}$$

In particular, this inequality holds for shared states $\rho$ where the subsystem of Alice is maximally mixed. In this case the Bloch vector of Alice is the zero vector $\boldsymbol{a} = \boldsymbol{0}$. Another scenario satisfying the inequality is given by the states with correlation tensor proportional to the identity matrix, i.e., $E_{ij} = \mu \delta_{ij}$. In this case the eigenvalues of $E E^T$ are all equal to $\mu^2$, and there is no further restriction on the Bloch vector of Alice.

An important family of states satisfying both of these conditions are the Werner states

$$\rho_w = p\,|\psi^-\rangle\,\langle\psi^-| + (1 - p)\frac{\mathbb{1}}{4} \tag{5.16}$$

with the singlet $|\psi^-\rangle = (|01\rangle - |10\rangle)/\sqrt{2}$. The state is separable for $p \leq 1/3$ which can be seen by checking the positivity of the partial transpose. As can also be seen by inspection, the elements of the correlation tensor are given as $E_{ij} = -p\delta_{ij}$, and the geometric measure of discord for this state becomes $D_G(\rho_w) = p^2/2$. In [1] these states were compared to another family of states given by

$$\sigma = \frac{1-k}{4}\,|\psi^+\rangle\,\langle\psi^+| + \frac{1+3k}{4}\,|\psi^-\rangle\,\langle\psi^-|$$
$$+ \frac{1-2t-k}{4}\,|00\rangle\,\langle00| + \frac{1+2t-k}{4}\,|11\rangle\,\langle11| \tag{5.17}$$

with $|\psi^+\rangle = (|01\rangle + |10\rangle)/\sqrt{2}$. For this family of states it can be verified by inspection that the elements of the correlation tensor are given as $E_{ij} = -k\delta_{ij}$, and thus the geometric measure of discord of this state is given by $D_G(\sigma) = k^2/2$. As was also noted in [1], for the parameters $k = 1/5$ and $t = 2/5$ the state is entangled. This can be verified by calculating the concurrence $C$ using the formula given in [6]: $C = 1/5$.

Combining the aforementioned results, we see that the average payoff for the separable state $\rho_w$ with $p = 1/3$ is bounded below as $\langle \mathcal{P}_{\max} \rangle \geq \min_{\boldsymbol{\beta}} \langle \mathcal{P}_{\max} \rangle = 2D_G(\rho_w) = 1/9$. In [1] this result was compared to the average payoff achievable with the entangled state $\sigma$ for the parameters $k = 1/5$ and $t = 2/5$: $\langle \mathcal{P}_{\max} \rangle \geq \min_{\boldsymbol{\beta}} \langle \mathcal{P}_{\max} \rangle = 2D_G(\sigma) = 1/25$. These results imply that for some directions $\boldsymbol{\beta}$ the separable state $\rho_w$ leads to a higher average payoff when compared to the state $\sigma$, despite the fact that the latter state is entangled.

### 5.1.5 Discussion

Dakić et al. conclude that a shared separable state can show a better performance for remote state preparation when compared to entangled states [1]. In particular, if Alice and Bob strictly follow the protocol, i.e., Alice performs von Neumann measurements and Bob conditionally applies a $\pi$ rotation around a given axis, there exist scenarios where shared entangled states can be outperformed by shared states without any entanglement. As a quantifier of the performance of the process Dakić et al. introduced a payoff function $\mathcal{P}$, and showed that the average optimal payoff is bounded below by the geometric measure of discord in a large number of scenarios. In these situations, the presence of discord guarantees that remote state preparation can always be achieved with nonzero average payoff. Experiment supporting these results has also been reported [1].

We complete the discussion by referring to the recent criticism of this approach. On the one hand, it was shown in [7] that a state can lead to nonzero average payoff even if its discord has been produced by local noise. According to [7] such states are unlikely to be useful in quantum information theory. On the other hand, the restriction of the protocol to von Neumann measurements of Alice and conditional rotations of Bob was criticized in [8]. This issue was further explored in [9], where it was shown that by relaxing these restrictions the advantage of separable states disappears if the standard fidelity $(1 + \mathbf{r} \cdot \mathbf{s})/2$ is used as a figure of merit of the protocol. However, regardless of this objection, it was also shown in [9] that in some situations separable states can still provide advantage for remote state preparation also for the standard fidelity $(1 + \mathbf{r} \cdot \mathbf{s})/2$.

## 5.2 Entanglement Distribution

### 5.2.1 General Protocol for Entanglement Distribution

The role of quantum discord in the task of *entanglement distribution* was considered in [10, 11]. The general setting is illustrated in Fig. 5.2: Alice is initially in possession of two particles, $A$ and $C$, while Bob is in possession of one particle $B$ (upper part of Fig. 5.2). If Alice sends the particle $C$ to Bob via a perfect quantum channel (middle part of Fig. 5.2), they end up in the final setup, where Bob is in possession of both particles $B$ and $C$, while Alice is in possession of $A$ (lower part of Fig. 5.2).

If the total state shared by Alice and Bob is $\rho = \rho^{ABC}$, then the initial amount of entanglement between them is given by $E^{AC|B} = E^{AC|B}(\rho)$, while the final amount of entanglement after sending the particle $C$ is given by $E^{A|BC} = E^{A|BC}(\rho)$. The amount of entanglement distributed in this process is then given by the difference between the final and the initial entanglement: $E^{A|BC} - E^{AC|B}$. In the following, the entanglement is quantified via the relative entropy of entanglement. For two parties $X$ and $Y$ it is defined as the minimal relative entropy between a given state $\rho^{XY}$ and

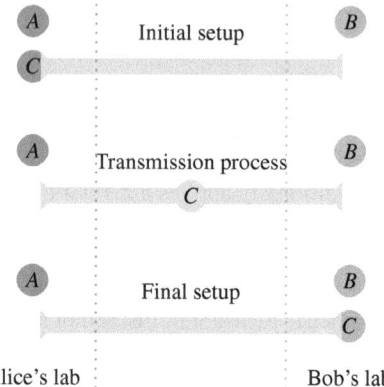

**Fig. 5.2** General protocol for entanglement distribution [10]. Copyright (2012) by the American Physical Society

the set of separable states $\mathcal{S}$:

$$E^{X|Y}(\rho^{XY}) = \min_{\sigma^{XY} \in \mathcal{S}} S(\rho^{XY}||\sigma^{XY}), \tag{5.18}$$

where $S(\rho||\sigma) = \mathrm{Tr}[\rho \log_2 \rho] - \mathrm{Tr}[\rho \log_2 \sigma]$ is the relative entropy between the states $\rho$ and $\sigma$.

### 5.2.2 The Role of Quantum Correlations

The main result of [10, 11] is the finding that the amount of entanglement $E^{A|BC} - E^{AC|B}$ distributed in this protocol is limited by the amount of discord between the exchanged particle $C$ and the rest of the system $AB$:

$$E^{A|BC} - E^{AC|B} \le \Delta^{C|AB}. \tag{5.19}$$

Here, $\Delta^{C|AB} = \Delta^{C|AB}(\rho)$ is the relative entropy of discord, defined as the minimal relative entropy between a given state $\rho^{XY}$ and the same state after a local von Neumann measurement:

$$\Delta^{X|Y}(\rho^{XY}) = \min_{\{\Pi_i^X\}} S(\rho^{XY}||\sum_i \Pi_i^X \rho^{XY} \Pi_i^X), \tag{5.20}$$

and the minimum is taken over local von Neumann measurements $\{\Pi_i^X\}$ on the subsystem $X$.

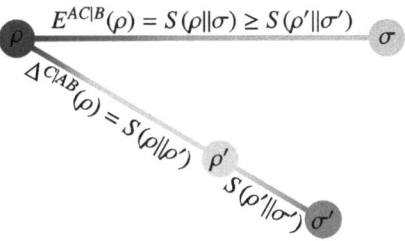

$$E^{AC|B}(\rho) = S(\rho\|\sigma) \geq S(\rho'\|\sigma')$$

$$\Delta^{C|AB}(\rho) = S(\rho\|\rho')$$

$$S(\rho'\|\sigma')$$

**Fig. 5.3** Proof of the main result in Eq. (5.19) [10]. Copyright (2012) by the American Physical Society

In the following we will reproduce the proof of Eq. (5.19) as presented in [10] and sketched in Fig. 5.3. In particular, consider the state $\sigma$ which is separable with respect to the bipartition $AC|B$ and at the same time the closest separable state to $\rho$, i.e., $E^{AC|B}(\rho) = S(\rho\|\sigma)$. Moreover, define a von Neumann measurement $\{\Pi_i^C\}$ such that the measured state $\rho' = \sum_i \Pi_i^C \rho \Pi_i^C$ has minimal relative entropy to $\rho$, i.e., $\Delta^{C|AB}(\rho) = S(\rho\|\rho')$. Finally, the state $\sigma' = \sum_i \Pi_i^C \sigma \Pi_i^C$ is defined by applying the same von Neumann measurement on the state $\sigma$. The proof of Eq. (5.19) now follows by observing that the states $\rho$, $\rho'$ and $\sigma'$ lie on a straight line [10]:

$$S(\rho\|\sigma') = S(\rho\|\rho') + S(\rho'\|\sigma'). \tag{5.21}$$

Before we prove this equality, we note that all quantities in this expression are finite. In particular, $S(\rho\|\rho')$ is finite due to the definition of the state $\rho'$: $S(\rho\|\rho') = \Delta^{C|AB}(\rho)$. This also implies that the support of $\rho$ is contained in the support of $\rho'$. Moreover, $S(\rho'\|\sigma')$ is finite due to the fact that the relative entropy does not increase under quantum operations, and thus

$$S(\rho'\|\sigma') \leq S(\rho\|\sigma) = E^{AC|B}(\rho). \tag{5.22}$$

This also means that the support of $\rho'$ is contained in the support of $\sigma'$. Combining these results we see that the support of $\rho$ is contained in the support of $\sigma'$, and thus $S(\rho\|\sigma')$ is also finite.

For proving Eq. (5.21) we will first prove the following equalities:

$$\mathrm{Tr}[\rho \log_2 \sigma'] = \mathrm{Tr}[\rho' \log_2 \sigma'], \tag{5.23a}$$

$$\mathrm{Tr}[\rho \log_2 \rho'] = \mathrm{Tr}[\rho' \log_2 \rho'], \tag{5.23b}$$

where all quantities are finite due to the arguments mentioned above. Eq. (5.23a) can be proven by noting that the state $\sigma' = \sum_i \Pi_i^C \sigma \Pi_i^C$ has the form of a quantum-classical state: $\sigma' = \sum_i p_i \sigma_i^{AB} \otimes \Pi_i^C$ with positive probabilities $p_i > 0$. If we further express the states $\sigma_i^{AB}$ in the eigendecomposition $\sigma_i^{AB} = \sum_j \lambda_{ij} |\psi_{ij}\rangle \langle\psi_{ij}|^{AB}$ with positive eigenvalues $\lambda_{ij} > 0$ and eigenstates $|\psi_{ij}^{AB}\rangle$, we obtain the following:

$$\text{Tr}[\rho' \log_2 \sigma'] = \text{Tr}\left[\left(\sum_k \Pi_k^C \rho \Pi_k^C\right)\left(\sum_{ij} \log_2(p_i \lambda_{ij}) |\psi_{ij}\rangle \langle\psi_{ij}|^{AB} \otimes \Pi_i^C\right)\right] \quad (5.24a)$$

$$= \sum_{ijk} \log_2(p_i \lambda_{ij})\text{Tr}\left[\Pi_k^C \rho \Pi_k^C |\psi_{ij}\rangle \langle\psi_{ij}|^{AB} \otimes \Pi_i^C\right] \quad (5.24b)$$

$$= \sum_{ijk} \log_2(p_i \lambda_{ij})\text{Tr}\left[\rho |\psi_{ij}\rangle \langle\psi_{ij}|^{AB} \otimes \Pi_k^C \Pi_i^C \Pi_k^C\right] \quad (5.24c)$$

$$= \sum_{ij} \log_2(p_i \lambda_{ij})\text{Tr}\left[\rho |\psi_{ij}\rangle \langle\psi_{ij}|^{AB} \otimes \Pi_i^C\right] \quad (5.24d)$$

$$= \text{Tr}\left[\rho \sum_{ij} \log_2(p_i \lambda_{ij}) |\psi_{ij}\rangle \langle\psi_{ij}|^{AB} \otimes \Pi_i^C\right] = \text{Tr}[\rho \log_2 \sigma']. \quad (5.24e)$$

In Eq. (5.24b) we used the linearity of the trace, and in Eq. (5.24c) its cyclic invariance. In Eq. (5.24d) we used the orthogonality of projectors, i.e., $\Pi_k^C \Pi_i^C \Pi_k^C = \delta_{ki} \Pi_i^C$. By applying the linearity of the trace once again in Eq. (5.24e) we arrive at the desired result: $\text{Tr}[\rho \log_2 \sigma'] = \text{Tr}[\rho' \log_2 \sigma']$. Using the same arguments Eq. (5.23b) is also seen to be correct.

The proof of Eq. (5.21) now follows by applying these results to the sum $S(\rho||\rho') + S(\rho'||\sigma')$:

$$S(\rho||\rho') + S(\rho'||\sigma') = \text{Tr}[\rho \log_2 \rho] - \text{Tr}[\rho \log_2 \rho'] + \text{Tr}[\rho' \log_2 \rho'] - \text{Tr}[\rho' \log_2 \sigma']$$

$$\overset{\text{Eq. (5.23b)}}{=} \text{Tr}[\rho \log_2 \rho] - \text{Tr}[\rho \log_2 \rho'] + \text{Tr}[\rho' \log_2 \rho'] - \text{Tr}[\rho' \log_2 \sigma']$$

$$= \text{Tr}[\rho \log_2 \rho] - \text{Tr}[\rho' \log_2 \sigma']$$

$$\overset{\text{Eq. (5.23a)}}{=} \text{Tr}[\rho \log_2 \rho] - \text{Tr}[\rho \log_2 \sigma'] = S(\rho||\sigma'). \quad (5.25)$$

We now turn to the proof of the main result in Eq. (5.19). Starting from Eq. (5.21) and recalling that the state $\rho'$ was defined such that $S(\rho||\rho') = \Delta^{C|AB}(\rho)$ we obtain the following equality: $S(\rho||\sigma') = \Delta^{C|AB}(\rho) + S(\rho'||\sigma')$. In the next step we make use of Eq. (5.22) arriving at the following result:

$$S(\rho||\sigma') \leq \Delta^{C|AB}(\rho) + E^{AC|B}(\rho). \quad (5.26)$$

In the final step, recall that the state $\sigma$ is separable with respect to the bipartition $AC|B$, and thus can be written as $\sigma = \sum_j q_j \sigma_j^{AC} \otimes \sigma_j^B$. Using this expression, we can write the state $\sigma' = \sum_i \Pi_i^C \sigma \Pi_i^C$ as

$$\sigma' = \sum_{ij} p_{ij} q_j \sigma_{ij}^A \otimes \sigma_j^B \otimes \Pi_i^C \quad (5.27)$$

with $\sum_i \Pi_i^C \sigma_j^{AC} \Pi_i^C = \sum_i p_{ij} \sigma_{ij}^A \otimes \Pi_i^C$. From this result we see that the state $\sigma'$ is fully separable, and thus the relative entropy between $\rho$ and $\sigma'$ is an upper bound on the relative entropy of entanglement $E^{A|BC}$, i.e., $E^{A|BC}(\rho) \le S(\rho||\sigma')$. Inserting this inequality into Eq. (5.26) we arrive at the desired result: $E^{A|BC}(\rho) \le \Delta^{C|AB}(\rho) + E^{AC|B}(\rho)$. This completes the proof of Eq. (5.19).

As was further pointed out in [10], the inequality (5.19) also holds in a more general case, where the relative entropy $S$ in both Eqs. (5.18) and (5.20) is replaced by a general distance $D$ which has the following properties:

- $D$ does not increase under quantum operations, i.e.,

$$D(\Lambda[\rho], \Lambda[\sigma]) \le D(\rho, \sigma) \qquad (5.28)$$

for any quantum operation $\Lambda$ and any pair of states $\rho$ and $\sigma$,
- $D$ satisfies the triangle inequality, i.e.,

$$D(\rho, \sigma) \le D(\rho, \tau) + D(\tau, \sigma) \qquad (5.29)$$

for any states $\rho$, $\sigma$, and $\tau$.

By virtue of the triangle inequality, Eq. (5.21) changes to the inequality

$$D(\rho, \sigma') \le D(\rho, \rho') + D(\rho', \sigma'). \qquad (5.30)$$

Starting from this result, Eq. (5.19) can be proven following the same reasoning as for the relative entropy. Important examples for distances having these properties are the trace distance $D_t(\rho_1, \rho_2) = \frac{1}{2}\mathrm{Tr}|\rho_1 - \rho_2|$ with the trace norm of an operator $M$ defined as $\mathrm{Tr}|M| = \mathrm{Tr}\sqrt{M^\dagger M}$ and the Bures distance $D_B(\rho_1, \rho_2) = 2(1 - \sqrt{F(\rho_1, \rho_2)})$ with the fidelity $F(\rho_1, \rho_2) = \left(\mathrm{Tr}\sqrt{\sqrt{\rho_1}\rho_2\sqrt{\rho_1}}\right)^2$.

### 5.2.3 Discussion

The result in Eq. (5.19) reveals a fundamental relation between the amount of entanglement in different splits of a quantum system. This can be seen by permuting the parties $A$ and $B$ in Eq. (5.19), which delivers the following inequality [11]:

$$\left| E^{A|BC} - E^{AC|B} \right| \le \Delta^{C|AB}. \qquad (5.31)$$

This inequality provides a strong link between $E^{A|BC}$ and $E^{AC|B}$. In particular, for zero quantum discord $\Delta^{C|AB} = 0$ this result immediately implies that these quantities are equal: $E^{A|BC} = E^{AC|B}$.

Finally, we notice that for successful entanglement distribution the exchanged particle does not need to be entangled with the rest of the system. In particular, there exist states $\rho = \rho^{ABC}$ which exhibit no entanglement between the exchanged particle $C$ and the rest of the system $AB$, i.e,

$$\rho = \sum_i p_i \cdot \rho_i^{AB} \otimes \rho_i^C, \qquad (5.32)$$

while at the same time the final amount of entanglement $E^{A|BC}$ is strictly larger than the initial amount of entanglement $E^{AC|B}$: $E^{A|BC}(\rho) > E^{AC|B}(\rho)$. The possibility of such *entanglement distribution with separable states* was first pointed out by Cubitt et al. in [12], who proved that this phenomenon is possible with vanishing initial entanglement $E^{AC|B}(\rho) = 0$. In the last years these results were further extended to different classes of quantum states [11, 13–15], and a classical counterpart for this quantum phenomenon was also presented [16]. The limits for this effect were further explored in [17], where it was shown that entanglement distribution with separable states requires states with rank at least three if the amount of entanglement is quantified via the logarithmic negativity. Very recently, three independent experiments have also shown that entanglement distribution with separable states is indeed possible with current technology [18–21].

## 5.3 Transmission of Correlations

### 5.3.1 Classical Transmission of Correlations

The role of quantum discord for the transmission of correlations was studied in [22]. The setup is illustrated in Fig. 5.4: initially Alice and Bob share a joint quantum state $\rho^{AB}$. The third party Charlie is initially uncorrelated with Alice and Bob, i.e., the total initial state is given by

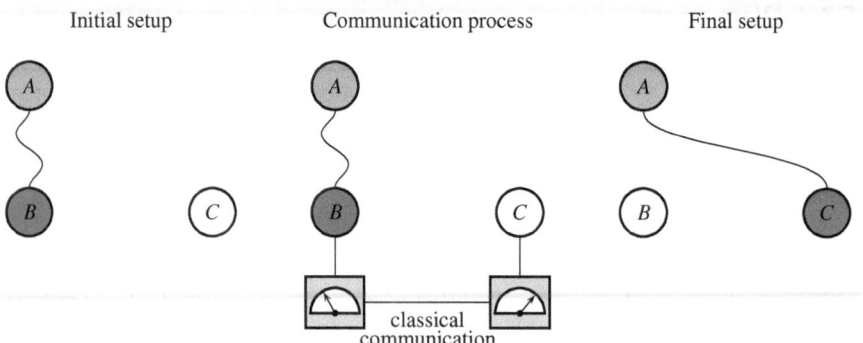

**Fig. 5.4** Classical transmission of correlations

$$\rho^{ABC} = \rho^{AB} \otimes \rho^{C}, \qquad (5.33)$$

see left part of Fig. 5.4. In the task of *classical transmission* Bob aims to transfer his state to Charlie by the means of local operations and classical communication (LOCC), see middle part of Fig. 5.4. After this process the final state takes the form

$$\rho_f^{ABC} = \Lambda_{B \leftrightarrow C}[\rho^{ABC}], \qquad (5.34)$$

where $\Lambda_{B \leftrightarrow C}$ denotes an LOCC operation between Bob and Charlie. In the ideal case the final state $\rho_f^{AC}$ shared by Alice and Charlie is equal to the initial state $\rho^{AB}$ shared by Alice and Bob (right part of Fig. 5.4):

$$\rho_f^{AC} = \rho^{AB}. \qquad (5.35)$$

As was pointed out in [22], such an ideal process is not possible in general. In particular, due to the fact that entanglement cannot be created by LOCC, the final state $\rho_f^{AC}$ is always separable. This implies that Eq. (5.35) is never fulfilled if Alice and Bob share an entangled initial state $\rho^{AB}$. As was further shown in [22], the ideal process is possible if and only if Alice and Bob share a quantum-classical state:

$$\rho^{AB} = \sum_i p_i \rho_i^A \otimes |i\rangle \langle i|^B. \qquad (5.36)$$

This was shown by introducing a figure of merit $I^c$ which quantifies the maximal mutual information between Alice and Charlie achievable in this procedure. The formal definition of $I^c$ can be given as follows:

$$I^c(\rho^{AB}) = \lim_{d_C \to \infty} \sup_{\Lambda_{B \leftrightarrow C}} I(\rho_f^{AC}), \qquad (5.37)$$

where the supremum is taken over all LOCC operations $\Lambda_{B \leftrightarrow C}$ between Bob and Charlie, $I$ is the mutual information, and $d_C$ is the dimension of Charlie's system.

### 5.3.2 The Role of Quantum Correlations

As was shown in [22], the figure of merit for the classical transmission $I^c$ introduced in Eq. (5.37) is closely related to the amount of discord in the initial state $\rho^{AB}$. The latter is defined as

$$D^{B|A}(\rho^{AB}) = I(\rho^{AB}) - \sup_{\{M_i^B\}} J(\rho^{AB})_{\{M_i^B\}}, \qquad (5.38)$$

where $J(\rho^{AB})_{\{M_i^B\}}$ is given as

$$J(\rho^{AB})_{\{M_i^B\}} = S(\rho^A) - \sum_i p_i S(\rho_i^A). \tag{5.39}$$

Here, $\{M_i^B\}$ is a positive operator-valued measure (POVM) on Bob's system $B$, $p_i = \mathrm{Tr}[M_i^B \rho^{AB}]$ is the probability for the outcome $i$, and $\rho_i^A = \mathrm{Tr}_B[M_i^B \rho^{AB}]/p_i$ is the state of Alice after the outcome $i$ has been obtained. $I^c$ is related to the discord $D^{B|A}$ as follows:

$$I^c(\rho^{AB}) = I(\rho^{AB}) - D^{B|A}(\rho^{AB}). \tag{5.40}$$

This equality was proven in [22], and we will present an alternative proof in the following. In particular we will prove the inequalities

$$I^c(\rho^{AB}) \leq I(\rho^{AB}) - D^{B|A}(\rho^{AB}), \tag{5.41}$$

$$I^c(\rho^{AB}) \geq I(\rho^{AB}) - D^{B|A}(\rho^{AB}), \tag{5.42}$$

which taken together imply Eq. (5.40).

For proving Eq. (5.41) we consider the structure of the final state $\rho_f^{ABC} = \Lambda_{B \leftrightarrow C}[\rho^{ABC}]$ using the fact that any LOCC operation $\Lambda_{B \leftrightarrow C}$ can be written as a separable operation[1]

$$\rho_f^{ABC} = \Lambda_{B \leftrightarrow C}[\rho^{ABC}] = \sum_{i=1}^{m} B_i \otimes C_i \rho^{ABC} B_i^\dagger \otimes C_i^\dagger \tag{5.43}$$

with a finite number of terms $m$ and Kraus operators $B_i \otimes C_i$ satisfying $\sum_{i=1}^{m} B_i^\dagger B_i \otimes C_i^\dagger C_i = \mathbb{1}^B \otimes \mathbb{1}^C$ [23]. In the next step recall that the initial state $\rho^{ABC}$ has the form $\rho^{ABC} = \rho^{AB} \otimes \rho^C$. Moreover, $I^c$ does not depend on the choice of the state $\rho^C$, and thus we choose $\rho^C = \mathbb{1}^C/d_C$. With this in mind, the final state $\rho_f^{ABC}$ can also be written as

$$\rho_f^{ABC} = \frac{1}{d_C} \sum_{i=1}^{m} B_i \rho^{AB} B_i^\dagger \otimes C_i C_i^\dagger. \tag{5.44}$$

Now we define positive numbers $q_i = \mathrm{Tr}[C_i C_i^\dagger] > 0$ and quantum states $\sigma_i^C = C_i C_i^\dagger/q_i$. The expression for the final state $\rho_f^{ABC}$ further reduces to

$$\rho_f^{ABC} = \sum_{i=1}^{m} E_i^B \rho^{AB} \left(E_i^B\right)^\dagger \otimes \sigma_i^C. \tag{5.45}$$

---

[1] The inverse is not true in general, i.e., a separable operation does not necessarily correspond to LOCC [23].

Here, $E_i^B$ are Kraus operators on the subsystem $B$ defined as $E_i^B = \sqrt{\frac{q_i}{d_C}} B_i$. The fact that $E_i^B$ are indeed Kraus operators, i.e., satisfy $\sum_{i=1}^{m} \left( E_i^B \right)^\dagger E_i^B = \mathbb{1}^B$, can be verified by inspection:

$$\sum_{i=1}^{m} \left( E_i^B \right)^\dagger E_i^B = \sum_{i=1}^{m} \frac{q_i}{d_C} B_i^\dagger B_i = \frac{1}{d_C} \sum_{i=1}^{m} \mathrm{Tr}[C_i C_i^\dagger] \cdot B_i^\dagger B_i \qquad (5.46)$$

$$= \frac{1}{d_C} \mathrm{Tr}_C \left[ \sum_{i=1}^{m} B_i^\dagger B_i \otimes C_i^\dagger C_i \right] = \frac{1}{d_C} \mathrm{Tr}_C \left[ \mathbb{1}^B \otimes \mathbb{1}^C \right] = \mathbb{1}^B.$$

Starting from the result in Eq. (5.45) the final state shared by Alice and Charlie takes the form

$$\rho_f^{AC} = \sum_{i=1}^{m} \mathrm{Tr}_B \left[ M_i^B \rho^{AB} \right] \otimes \sigma_i^C, \qquad (5.47)$$

where $M_i^B$ are POVM elements on the subsystem $B$ defined as $M_i^B = \left( E_i^B \right)^\dagger E_i^B$. We will now show that for any such state the mutual information is bounded above as follows:

$$I(\rho_f^{AC}) \leq I(\rho^{AB}) - D^{B|A}(\rho^{AB}). \qquad (5.48)$$

Since the figure of merit $I^c$ was defined as the supremum of the mutual information between Alice and Charlie over all LOCC protocols in the limit $d_C \to \infty$, this result will imply the inequality (5.41). To prove this statement we introduce the state

$$\tau^{A\tilde{C}} = \sum_{i=1}^{m} \mathrm{Tr}_B \left[ M_i^B \rho^{AB} \right] \otimes |i\rangle \langle i|^{\tilde{C}} \qquad (5.49)$$

with a new system $\tilde{C}$ having dimension $d_{\tilde{C}} = \max\{d_C, m\}$. Note that the latter state can be transformed into the state $\Lambda_{\tilde{C}}[\tau^{A\tilde{C}}] = \sum_{i=1}^{m} \mathrm{Tr}_B \left[ M_i^B \rho^{AB} \right] \otimes \sigma_i^{\tilde{C}}$ by a local operation[2] $\Lambda_{\tilde{C}}$, where the states $\sigma_i^{\tilde{C}}$ are the same as $\sigma_i^C$ in Eq. (5.47). Since the mutual information does not increase under local operations, it follows that the state $\tau^{A\tilde{C}}$ has at least the same mutual information as $\rho_f^{AC}$:

$$I(\tau^{A\tilde{C}}) \geq I(\rho_f^{AC}). \qquad (5.50)$$

---

[2] The local operation that achieves this task is a measure-and-prepare map with Kraus operators $K_{ab}^{\tilde{C}} = \sqrt{\sigma_b^{\tilde{C}}} |a\rangle \langle b|^{\tilde{C}}$.

Finally, it is straightforward to verify that the mutual information of $\tau^{A\tilde{C}}$ can be written as

$$I(\tau^{A\tilde{C}}) = J(\rho^{AB})_{\{M_i^B\}} \tag{5.51}$$

with $J(\rho^{AB})_{\{M_i^B\}}$ defined in Eq. (5.39). Combining these results we arrive at the inequality

$$I(\rho_f^{AC}) \le I(\tau^{A\tilde{C}}) \le \sup_{\{M_i^B\}} J(\rho^{AB})_{\{M_i^B\}} = I(\rho^{AB}) - D^{B|A}(\rho^{AB}), \tag{5.52}$$

where the last equality follows from the definition of discord in Eq. (5.38). This completes the proof of Eq. (5.48) and the inequality (5.41).

We will now complete the proof of Eq. (5.40) by proving the inequality (5.42). This can be done by considering a specific LOCC protocol where Bob performs a measurement with $d_C$ Kraus operators $E_i^B$ on his subsystem $B$. The outcome of the measurement is sent to Charlie who stores it in his system $C$ of dimension $d_C$. After performing this protocol, the final state $\rho_f^{AC}$ shared by Alice and Charlie takes the form

$$\rho_f^{AC} = \sum_{i=1}^{d_C} \mathrm{Tr}_B \left[ M_i^B \rho^{AB} \right] \otimes |i\rangle \langle i|^C \tag{5.53}$$

with POVM elements $M_i^B = \left( E_i^B \right)^\dagger E_i^B$. Since we consider a specific LOCC protocol, $I_c$ cannot be smaller than the mutual information for any state obtained in this way, and thus

$$I^c(\rho^{AB}) \ge \lim_{d_C \to \infty} \sup_{\{M_i^B\}} I(\rho_f^{AC}). \tag{5.54}$$

The inequality (5.42) follows by noting that the state $\rho_f^{AC}$ has the same form as $\tau^{A\tilde{C}}$ in Eq. (5.49), and thus by applying the same arguments as for $\tau^{A\tilde{C}}$ we see that the mutual information of $\rho_f^{AC}$ can be written as $I(\rho_f^{AC}) = J(\rho^{AB})_{\{M_i^B\}}$. Together with the definition of discord in Eq. (5.38) this completes the proof of Eqs. (5.42) and (5.40).

### 5.3.3 Quantum Transmission of Correlations

The task of *quantum transmission* was considered in [22] and independently in [24]. The setup is illustrated in Fig. 5.5. Similar to the scenario for the classical transmission, Alice and Bob share a joint initial state $\rho^{AB}$ (left part of Fig. 5.5). The system of Charlie now consists of two subsystems $C_1$ and $C_2$, initially uncorrelated

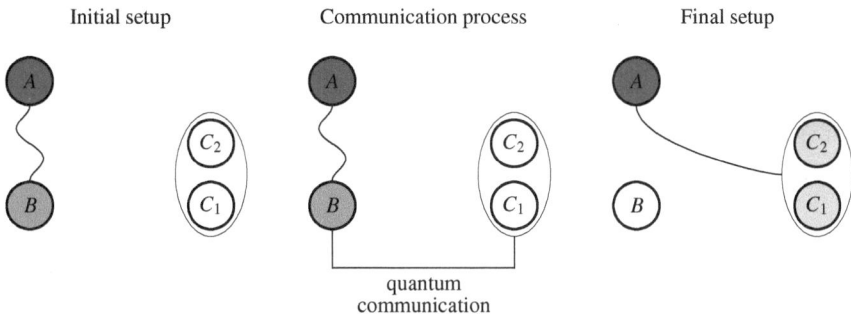

**Fig. 5.5** Quantum transmission of correlations

with Alice and Bob, and the total initial state is thus given by

$$\rho^{ABC} = \rho^{AB} \otimes \rho^{C_1 C_2}. \tag{5.55}$$

Moreover, Bob and Charlie have access to a general quantum communication channel $\Lambda_{BC}$ (middle part of Fig. 5.5), and the final state after the application of the channel takes the form

$$\rho_f^{ABC} = \Lambda_{BC}[\rho^{ABC}], \tag{5.56}$$

see right part of Fig. 5.5.

The aim of this process is to achieve maximal mutual information between the system of Alice and each of Charlie's subsystems $C_1$ and $C_2$ on average. Following [22] we denote the corresponding figure of merit by $I_2^q$, where the superscript $q$ tells us that quantum communication is considered, and the index 2 gives the number of Charlie's subsystems. The formal definition of $I_2^q$ can be given as follows:

$$I_2^q(\rho^{AB}) = \lim_{d\to\infty} \sup_{\Lambda_{BC}} \frac{I(\rho_f^{AC_1}) + I(\rho_f^{AC_2})}{2}, \tag{5.57}$$

where both subsystems of Charlie have the same dimension $d = d_{C_1} = d_{C_2}$. It is straightforward to generalize this quantity to $n$ subsystems of Charlie:

$$I_n^q(\rho^{AB}) = \lim_{d\to\infty} \sup_{\Lambda_{BC}} \frac{\sum_{i=1}^n I(\rho_f^{AC_i})}{n}, \tag{5.58}$$

where $d = d_{C_1} = d_{C_2} = \ldots = d_{C_n}$ is the dimension of each of Charlie's subsystems.

## 5.3.4 Equivalence of Quantum and Classical Transmission
### for Pure States

In the scenario where Alice and Bob share a pure initial state $\rho^{AB} = |\psi\rangle\langle\psi|^{AB}$ quantum and classical transmission are equivalent [22]:

$$I_n^q(|\psi\rangle\langle\psi|^{AB}) = I^c(|\psi\rangle\langle\psi|^{AB}) \tag{5.59}$$

for any number of Charlie's subsystems $n \geq 2$. We will present the proof for this statement following the arguments of [22]. First, it is important to note that $I_n^q$ cannot be smaller than $I^c$:

$$I_n^q(\rho^{AB}) \geq I^c(\rho^{AB}), \tag{5.60}$$

which follows from the fact that quantum communication is more general than classical communication. In the following we will show that for pure states $|\psi\rangle^{AB}$ and $n \geq 2$ the inverse inequality also holds:

$$I_n^q(|\psi\rangle\langle\psi|^{AB}) \leq I^c(|\psi\rangle\langle\psi|^{AB}). \tag{5.61}$$

This result together with Eq. (5.60) will complete the proof of the desired equality (5.59).

In the first step we will show that the sum $\sum_{i=1}^{n} I(\rho_f^{AC_i})$ is in general bounded above as follows:

$$\sum_{i=1}^{n} I(\rho_f^{AC_i}) \leq n S(\rho_f^A). \tag{5.62}$$

This inequality can be proven by using the fact that any tripartite state $\rho^{XYZ}$ satisfies the inequality

$$I(\rho^{XY}) + I(\rho^{XZ}) \leq 2S(\rho^X), \tag{5.63}$$

which can be seen by rewriting it as $S(\rho^Y) + S(\rho^Z) \leq S(\rho^{XY}) + S(\rho^{XZ})$ and noting that the latter inequality is equivalent to the strong subadditivity of the von Neumann entropy [see p. 521 in [25]]. If we apply this inequality to the state $\rho_f^{AC_kC_l}$ with $k \neq l$, we arrive at the following inequality:

$$I(\rho_f^{AC_k}) + I(\rho_f^{AC_l}) \leq 2S(\rho_f^A). \tag{5.64}$$

Starting from this result we will now prove Eq. (5.62) for even $n \geq 2$, i.e., $n = 2m$. In this case the sum can be bounded as

$$\sum_{i=1}^{n} I(\rho_f^{AC_i}) = \sum_{j=1}^{m} \left\{ I(\rho_f^{AC_{2j-1}}) + I(\rho_f^{AC_{2j}}) \right\} \leq 2mS(\rho_f^A) = nS(\rho_f^A), \quad (5.65)$$

where we used Eq. (5.64) to obtain $I(\rho_f^{AC_{2j-1}}) + I(\rho_f^{AC_{2j}}) \leq 2S(\rho_f^A)$. For odd $n \geq 3$ we can write

$$2\sum_{i=1}^{n} I(\rho_f^{AC_i}) = \sum_{k=1}^{n} I(\rho_f^{AC_k}) + \sum_{l=1}^{n} I(\rho_f^{AC_l})$$

$$= I(\rho_f^{AC_1}) + \sum_{k=2}^{n} I(\rho_f^{AC_k}) + \sum_{l=1}^{n-1} I(\rho_f^{AC_l}) + I(\rho_f^{AC_n})$$

$$= \sum_{k=2}^{n} I(\rho_f^{AC_k}) + \sum_{l=1}^{n-1} I(\rho_f^{AC_l}) + I(\rho_f^{AC_1}) + I(\rho_f^{AC_n}). \quad (5.66)$$

Note that the sums $\sum_{k=2}^{n} I(\rho_f^{AC_k})$ and $\sum_{l=1}^{n-1} I(\rho_f^{AC_l})$ each have $n-1$ terms, which is an even number. Thus, we can use the same arguments as in Eq. (5.65) to see that both of them are bounded from above by $(n-1)S(\rho_f^A)$. Finally, due to Eq. (5.64) the sum $I(\rho_f^{AC_1}) + I(\rho_f^{AC_n})$ is bounded from above by $2S(\rho_f^A)$. Combining these results we see that

$$2\sum_{i=1}^{n} I(\rho_f^{AC_i}) \leq 2nS(\rho_f^A), \quad (5.67)$$

which completes the proof of the desired inequality (5.62) for any $n \geq 2$.

Recalling that the state of Alice never changes in the process, i.e., $\rho_f^A = \rho^A$, Eq. (5.62) implies that the average mutual information $\frac{1}{n}\sum_{i=1}^{n} I(\rho_f^{AC_i})$ never exceeds the entropy of $\rho^A$:

$$\frac{1}{n}\sum_{i=1}^{n} I(\rho_f^{AC_i}) \leq S(\rho^A). \quad (5.68)$$

Due to the definition of $I_n^q$ in Eq. (5.58) this result immediately implies that $S(\rho^A)$ is also an upper bound for $I_n^q$:

$$I_n^q(\rho^{AB}) \leq S(\rho^A). \quad (5.69)$$

In the final step, note that for pure states $\rho^{AB} = |\psi\rangle\langle\psi|^{AB}$ the quantity $I^c$ coincides with the entropy of $\rho^A$:

$$I^c(|\psi\rangle\langle\psi|^{AB}) = S(\rho^A). \quad (5.70)$$

This follows from the relation between $I^c$ and quantum discord provided in Eq. (5.40) by noting that for pure states the mutual information and the discord are given as $I(|\psi\rangle\langle\psi|^{AB}) = 2S(\rho^A)$ and $D^{B|A}(|\psi\rangle\langle\psi|^{AB}) = S(\rho^A)$. This completes the proof of Eq. (5.61) and the desired equality (5.59) immediately follows.

### 5.3.5 Discussion

As was pointed out in [22], ideal classical transmission as illustrated in Fig. 5.4 is possible if and only if the initial state $\rho^{AB}$ is quantum-classical:

$$\rho^{AB} = \sum_i p_i \rho_i^A \otimes |i\rangle\langle i|^B . \tag{5.71}$$

For any other state the discord $D^{B|A}(\rho^{AB})$ is nonzero, and the process of classical transmission unavoidably leads to a loss of information, i.e., the mutual information between Alice and Charlie is never larger than the difference $I(\rho^{AB}) - D^{B|A}(\rho^{AB})$. The amount of quantum discord $D^{B|A}(\rho^{AB})$ thus quantifies the loss of information in the task of classical transmission.

It was further shown in [22] that the equivalence between quantum and classical transmission stated in Eq. (5.59) only holds for pure initial state. In particular, there exist mixed states $\rho^{AB}$ for which quantum transmission leads to a better performance when compared to the classical transmission: $I_2^q(\rho^{AB}) > I^c(\rho^{AB})$. In this context, an important result was obtained recently by Brandão et al., who showed that $I_n^q$ and $I^c$ coincide in the asymptotic limit [24]:

$$\lim_{n\to\infty} I_n^q(\rho^{AB}) = I^c(\rho^{AB}). \tag{5.72}$$

Thus, quantum and classical transmission are equivalent for any initial state $\rho^{AB}$ if the number of Charlie's subsystems goes to infinity.

### References

1. Dakić, B., et al.: Quantum discord as resource for remote state preparation. Nat. Phys. **8**, 666–670 (2012)
2. Bennett, C.H., et al.: Teleporting an unknown quantum state via dual classical and Einstein-Podolsky-Rosen channels. Phys. Rev. Lett. **70**, 1895–1899 (1993)
3. Bennett, C.H., et al.: Remote state preparation. Phys. Rev. Lett. **87**, 077902 (2001)
4. Horn, R.A., Johnson, C.R.: Matrix Analysis. Cambridge University Press, New York (1985)
5. Dakić, B., Vedral, V., Brukner, Č.: Necessary and sufficient condition for nonzero quantum discord. Phys. Rev. Lett. **105**, 190502 (2010)
6. Wootters, W.K.: Entanglement of formation of an arbitrary state of two qubits. Phys. Rev. Lett. **80**, 2245–2248 (1998)

7. Giorgi, G.L.: Quantum discord and remote state preparation. Phys. Rev. A **88**, 022315 (2013)
8. Tufarelli, T., Girolami, D., Vasile, R., Bose, S., Adesso, G.: Quantum resources for hybrid communication via qubit-oscillator states. Phys. Rev. A **86**, 052326 (2012)
9. Horodecki, P., Tuziemski, J., Mazurek, P., Horodecki, R.: Can communication power of separable correlations exceed that of entanglement resource? Phys. Rev. Lett. **112**, 140507 (2014)
10. Streltsov, A., Kampermann, H., Bruß, D.: Quantum cost for sending entanglement. Phys. Rev. Lett. **108**, 250501 (2012)
11. Chuan, T.K., et al.: Quantum discord bounds the amount of distributed entanglement. Phys. Rev. Lett. **109**, 070501 (2012)
12. Cubitt, T.S., Verstraete, F., Dür, W., Cirac, J.I.: Separable states can be used to distribute entanglement. Phys. Rev. Lett. **91**, 037902 (2003)
13. Mišta, L., Korolkova, N.: Distribution of continuous-variable entanglement by separable gaussian states. Phys. Rev. A **77**, 050302 (2008)
14. Kay, A.: Using separable bell-diagonal states to distribute entanglement. Phys. Rev. Lett. **109**, 080503 (2012)
15. Park, J., Lee, S.: Separable states to distribute entanglement. Int. J. Theor. Phys. **51**, 1100–1110 (2012)
16. Bae, J., Cubitt, T., Acín, A.: Nonsecret correlations can be used to distribute secrecy. Phys. Rev. A **79**, 032304 (2009)
17. Streltsov, A., Kampermann, H., Bruß, D.: Limits for entanglement distribution with separable states. Phys. Rev. A **90**, 032323 (2014)
18. Fedrizzi, A., et al.: Experimental distribution of entanglement with separable carriers. Phys. Rev. Lett. **111**, 230504 (2013)
19. Vollmer, C.E., et al.: Experimental entanglement distribution by separable states. Phys. Rev. Lett. **111**, 230505 (2013)
20. Peuntinger, C., et al.: Distributing entanglement with separable states. Phys. Rev. Lett. **111**, 230506 (2013)
21. Silberhorn, C.: Sharing entanglement without sending it. Physics **6**, 132 (2013)
22. Streltsov, A., Zurek, W.H.: Quantum discord cannot be shared. Phys. Rev. Lett. **111**, 040401 (2013)
23. Horodecki, R., Horodecki, P., Horodecki, M., Horodecki, K.: Quantum entanglement. Rev. Mod. Phys. **81**, 865–942 (2009)
24. Brandão, F.G.S.L., Piani, M., Horodecki, P.: Quantum Darwinism is Generic (2013). arXiv:1310.8640v1
25. Nielsen, M.A., Chuang, I.L.: Quantum Computation and Quantum Information. Cambridge University Press, Cambridge (2000)

# Chapter 6
# Outlook

In this work, we discussed the role of quantum correlations beyond entanglement in three fundamental tasks in quantum information theory: remote state preparation [1], entanglement distribution [2, 3], and transmission of correlations [4, 5]. Although these tasks clearly demonstrate the relevance of quantum discord and general quantum correlations in quantum information theory, they cannot cover the whole range of applications of quantum correlations beyond entanglement that have been presented recently. In the following, we will give an outlook on some of the developments in this direction.

The role of quantum discord in *quantum metrology* was first investigated by Modi et al. [6], and important contributions in this direction were made recently by Girolami et al. in [7, 8]. In the scenario considered in [8], Alice and Bob share a bipartite state $\rho^{AB}$ undergoing a local unitary evolution $U_A = e^{-i\varphi H_A}$ on the subsystem of Alice with a nondegenerate Hamiltonian $H_A$. The final state $U_A \rho^{AB} U_A^\dagger$ is then used to estimate the unknown parameter $\varphi$. As was shown in [8], the parameter $\varphi$ can always be estimated with nonzero precision whenever the state $\rho^{AB}$ is not classical-quantum, i.e., not of the form $\rho^{AB} = \sum_i p_i |i\rangle \langle i|^A \otimes \rho_i^B$. The authors of [8] investigate this phenomenon by introducing a new quantifier of quantum correlations which they call *interferometric power*. They show that the interferometric power is able to capture the worst-case precision of the procedure, and conclude that the presence of discord in a quantum state guarantees its usefulness for quantum metrology. Experiment supporting these theoretical results has also been reported in [8].

A great amount of attention was also attracted by the relation between entanglement and discord in the quantum measurement process [9–11]. In particular, it was shown in [9, 10] that for performing a von Neumann measurement on one part of a composite quantum state $\rho^{AB}$, the creation of entanglement between the system and the measurement apparatus is unavoidable whenever the state has nonzero quantum discord. Recently, experimental demonstration of this effect has also been

© The Author(s) 2015
A. Streltsov, *Quantum Correlations Beyond Entanglement*,
SpringerBriefs in Physics, DOI 10.1007/978-3-319-09656-8_6

reported [12]. These results support the role of quantum discord and general quantum correlations for studying entanglement on the one hand, and for understanding phenomena which cannot be explained solely by the presence of entanglement on the other hand. In this context, useful results can be expected from the investigation of quantum discord in the framework of coherence, recently introduced by Baumgratz et al. [13]. The main aim of this research direction would be the unification of all three concepts: entanglement, quantum correlations beyond entanglement, and coherence. This research may further lead to the discovery of new tasks in quantum information theory which are not based on entanglement, and which require new types of quantum correlations to capture their performance.

# References

1. Dakić, B., et al.: Quantum discord as resource for remote state preparation. Nat. Phys. **8**, 666–670 (2012)
2. Streltsov, A., Kampermann, H., Bruß, D.: Quantum cost for sending entanglement. Phys. Rev. Lett. **108**, 250–501 (2012)
3. Chuan, T.K., et al.: Quantum discord bounds the amount of distributed entanglement. Phys. Rev. Lett. **109**, 070501 (2012)
4. Streltsov, A., Zurek, W.H.: Quantum discord cannot be shared. Phys. Rev. Lett. **111**, 040401 (2013)
5. Brandão, F.G.S.L., Piani, M., Horodecki, P.: Quantum Darwinism is Generic (2013). arXiv:1310.8640v1
6. Modi, K., Cable, H., Williamson, M., Vedral, V.: Quantum correlations in mixed-state metrology. Phys. Rev. X **1**, 021022 (2011)
7. Girolami, D., Tufarelli, T., Adesso, G.: Characterizing nonclassical correlations via local quantum uncertainty. Phys. Rev. Lett. **110**, 240402 (2013)
8. Girolami, D., et al.: Quantum discord determines the interferometric power of quantum states. Phys. Rev. Lett. **112**, 210401 (2014)
9. Streltsov, A., Kampermann, H., Bruß, D.: Linking quantum discord to entanglement in a measurement. Phys. Rev. Lett. **106**, 160401 (2011)
10. Piani, M., et al.: All nonclassical correlations can be activated into distillable entanglement. Phys. Rev. Lett. **106**, 220403 (2011)
11. Gharibian, S., Piani, M., Adesso, G., Calsamiglia, J., Horodecki, P.: Characterizing quantumness via entanglement creation. Int. J. Quantum. Inform. **09**, 1701–1713 (2011)
12. Adesso, G., D'Ambrosio, V., Nagali, E., Piani, M., Sciarrino, F.: Experimental entanglement activation from discord in a programmable quantum measurement. Phys. Rev. Lett. **112**, 140501 (2014)
13. Baumgratz, T., Cramer, M., Plenio, M. B.: Quantifying coherence. Phys. Rev. Lett. **113**, 140401 (2014)

# Index

**B**
Bures distance, 9

**C**
Classical correlations, 17, 19
Classical-quantum states, 17
Concurrence, 14

**D**
Density operator, 5
Discord, 18
Distillable entanglement, 13

**E**
Entanglement, 11
    measure, 13
    of formation, 14
Entanglement distribution, 29
    with separable states, 34

**F**
Fidelity, 9

**G**
Geometric measure of discord, 21

**H**
Hilbert-Schmidt distance, 10

**I**
Information deficit, 20

**K**
Koashi-Winter relation, 20
Kraus operators, 7

**L**
LOCC, 12

**M**
Measurement operator, 6
Mixed state, 5
Mutual information, 8

**P**
Partial trace, 7
POVM, 6
Projective measurement, 6
Pure state, 5

**Q**
Quantum
    correlations, 17
    measurement, 6
    metrology, 45
    operation, 7
Qubit, 5

**R**
Reduced density operator, 7
Relative entropy, 8
    of discord, 21
    of entanglement, 15
    of quantumness, 21

© The Author(s) 2015
A. Streltsov, *Quantum Correlations Beyond Entanglement*,
SpringerBriefs in Physics, DOI 10.1007/978-3-319-09656-8

Remote state preparation, 23

**S**
Separable state, 11
Shannon entropy, 8
Singlet, 11

**T**
Trace distance, 9

Transmission of correlations
    classical, 34
    quantum, 38

**V**
Von Neumann
    entropy, 8
    measurement, 6